About the Author

Jack Miller is a veteran of Radar Site duty. He was assigned to an isolated radar site at Armstrong, Ontario Canada and to a remote Radar site in the high desert of Arizona outside of Winslow. He performed the duties of Security Police at the locations. Some of the stories included are from his experience while other stories are from other officers and airmen assigned to locations throughout the world.

After retiring from the USAF, Jack worked in law enforcement until he retired in 2002. At that time he began writing about his experiences. Presently he has five novels published.

Cold War Warrior (ISBN 978-1-935051-02-3)

Operation Switch (ISBN 978-1-4563558-0-7)

The Master Cheat (ISBN 978-0-9727840-3-0)

The Defector (ISBN 978-1-934551-04-7)

The Medal (ISBN 978-1-934051-45-0)

There is more information about each book at, and they may be ordered through, **www.retafsa.com** Books so ordered are inscribed, personalized and are sold at a reduced rate.

THE PEACEKEEPERS

Not all heroes receive medals

By Jack Miller

With stories from members of the United States Air Force
Radar Site Veterans Association

This book is a work of non-fiction. Although an identified portion is imaginary, it is based on fact.

The names used are real names and every effort has been made to attribute the individual stories to the correct contributor.

Copyright 2012, Jack Miller

ISBN: 978-1-936759-09-5

First Edition

DEDICATION

This book is dedicated to the men and women of the United States Air Force. Especially those who were assigned to the Air Defense Command (ADC). Thousands of officer and enlisted personnel spent time at Aircraft Control and Warning sites and the upgraded Radar Squadron sites during the cold war. Many of these airmen were forced to live apart from their loved ones. They endured the unfriendly weather and the loneliness of isolated and/or remote assignments. Some fractured the boredom by doing things they might never have done elsewhere. Some activities were good, some bad, some funny and some not so funny. Though this all they maintained their loyalty and dedication to the prevention of WW III. The free people of the United States should say thank you many times over to those who watched the skies.

To the wives, sweethearts and families of these men and women who served, and who remained supportive during those trying times, we, the former Airmen of the ADC cannot say thank you enough for allowing us to be dedicated to our jobs and our Country. If not for the love and support you showed us we might have lost the cold war.

A special thanks goes to Steve Weatherly, Les Crine, Wayne Fitzgerald, and Tom Steeves, all Air Force veterans who edited this manuscript. A very special thank you must go to Paul Jones who without his help this book would never have been completed.

Sincere thanks must go to the members of the Air Force Radar Site Veterans for their service, and to those who provided the stories used in this book. Some of the stories have been edited for grammar, length, etc.

A special note to all. The proceeds of this book will be donated to the Radar Site Museum. If you are ever in the area of Bellefontaine, Ohio please stop by.

INTRODUCTION

The following is an imaginary scenario; however, it actually happened many times during the cold war.

While others slept, the pilot of the lead Soviet Bear bomber in the flight of two others, known as the TU-95, aimed his plane east towards Alaska to what could be the start WW III. He was thinking, *It won't be long now and we should be picked up on their radar. It happens every time I fly this mission. My control center expects it and so do I. Hell it is what I hope. I want them attentive to their duties. If not it will delay our detection and could very well result in someone higher up the American chain of command not having enough time to make an intelligent decision. I want him to see me and to send up the interceptors. I don't want him pushing a panic button that sets off their bombers to my home in Russia.*

It should be only minutes now before I see the American jet fighters on my radar screen. Then within minutes I will see the lumbering F-89 Scorpions, what they call their "J" model, with their two white GENIE, Air to nuclear missiles hanging under the wings. They think those weapons are secrets, the fools. We know all about them. If ever they fired even one of the missiles at us, it would mean my instant death, the instant death of every crewman in my airplane and the others with me. The Americans who fire that missile probably don't even realize they might die too. Maybe I'll get lucky and see some of the new delta winged interceptor aircraft this trip. If I remember right they are called the F-102. They think we don't know about them either. Well my job is to test the American defenses, maybe find a weak spot, push them to their limit, and then hopefully I can turn my flight around and go home to the

USSR. I do not want to be the cause of a world war but I must follow my orders.

Ivan, as all Russian pilots were called by the Americans, again adjusted the two contra-rotating propellers of engine number four and he jotted a note for his ground mechanic to check the pitch of all eight blades. His adjustment tuned his variable pitch propellers and steadied the flight. He looked left and right and saw the other two Russian Bear aircraft, each with their crews of eight, right where they were supposed to be. Cruising at their normal speed of 435 kilometers per hour following their flight plans.

Ivan again thought about the possibility that some American might be asleep and get startled awake finding his aircraft well inside the Air Defense Identification Zone extending 200 miles out to sea from the U.S. coastline. If that happened they might shoot first and try to apologize later. He was hoping that the people sitting in front of the radar consoles were alert. He knew that any person watching one of the consoles might get bored or bleary-eyed and miss his echo. But he did not feel like dying that day so he was hoping they would detect him so he could return home early. He did not know where exactly these radar sites with their consoles were located except that they had to be built on high places; some so high even eagles wore oxygen masks.

As he neared his coast before flying over the emptiness of the waters between his country and Alaska he pressed his microphone button on the yoke of the plane and told his radar operator to be alert. They were approaching the edge of the American radar and soon he knew they would be painting on the radar consoles. He knew that if he was detected as he hoped, they would be scrambling interceptors to identify him. Then the fun would begin.

Tin City Air Force Station (AFS) used the call sign Dragnet. They and several other radar stations reported to the radar center located at Fire Island Alaska.

Fire Island AFS used the call sign of Slugger. They had direct lines to Elmendorf Air Force Base that housed the alert intercept aircraft of the Air Defense Command (ADC). The alert aircraft were F-89J's, called the Scorpion and made by Northrop Aircraft. The Scorpion was an all weather aircraft and ideal for duty in Alaska. It had a range of 1367 miles and a top speed of 636 miles per hour. Other models of the F-89 could carry a huge payload of bombs and rockets. Only the F-89J was assigned duty as "Alert" intercept aircraft, and when on alert status, it carried only two nuclear tipped genie missiles. Pilots when assigned on alert referred to the duty as suicide by genie. If they ever had to launch a genie missile at a group of enemy aircraft bent on attacking the United States, the nuclear missile would explode near the flight of enemy aircraft destroying them. When it did explode, everything in the blast zone would be vaporized. That might include the planes, pilots and radar operators of the launching aircraft.

It was at 0330 (3:30 am) on the morning or January 12th, 1959 at the height of the Cold War. A Radar Operator assigned to operations at Tin City AFS, Alaska was on duty. The Airman sat at a radar console in the operations area of the station. He had been on duty since midnight although it didn't make that much difference. It was winter and the daylight hours were only about three hours long each day. He had been watching the radar sweep going round and round on the console and thinking of home and warmer climes. Suddenly, he sat up straight in the chair and leaned slightly forward. He thought he had seen three echoes appear at the outer edge of his monitor. The sweep revolved and his suspicions were confirmed. Definitely three yellowish dots shown brightly as the sweep passed over them. The airman notified the sergeant who

verified the sighting. He notified the Weapons Assignment
Officer who called Slugger.

At almost the same time as the first report came in to Slugger
from Dragnet, a second confirming report was received from
another radar site located near Unalakleet Alaska. This one
named Galena and used the call sign Windmill. With two
reports, the Weapons Controller at Slugger pressed the alert
button signaling Elmendorf Air Force Base.

In the alert center, tower, fire department and security office,
the Klaxon horn began to blare out its OOOUUUGAA!
OOOUUUGAA! OOOUUUGAA! The racket brought the
assigned alert ground crews and air crews up from their sleep
making them instantly awake. They had been sleeping fully
clothed so all they had to do was to jump into their boots and
coats. The enlisted ground crews on alert were up first and
running to their vehicle which would take them to the alert
shelters. The Pilot and Radar Officer who sat in the rear seat
had a few more things to put on, shoulder holster, survival
vests, and other emergency equipment. Then they also ran to
their vehicle and were about 60 seconds behind the enlisted
crews. An armed security guard standing a post near the alert
hangers had also received the alarm. He knew he would be
seeing the alert crews heading towards his post. He had all
ready checked their security badges and recognized them at the
start of his tour saving them precious seconds. The Security
Police guard recognized the troops and waived the vehicle and
all its occupants through his post. He would wait for the
second vehicle containing the officers who would be
following. The first vehicle pulled to a stop in front of and
between the two active alert hangers and another guard posted
there, who had also pre-checked identification and faces,
allowed them entry to the alert hangers. A crew chief and their
assistant each entered one of the two alert hangers through
personnel doors located in the front hanger doors. In each

hanger was an F-89J, cockpit canopies raised and slung under each wing was a white nuclear missile. Under each wing were large metal drip pans to catch leaking fuel from their topped off wing fuel tanks.

The large front and rear hanger doors were opened, ground power units started and the safety pins with their red streamers pulled all in about fifty seconds.

The second vehicle containing the four officers arrived and slid the vehicle to a stop in the rear of the alert hangers. Another armed security guard posted there who also knew the crewmen watched the activity. One officer climbed the ladder and sat in the front seat. The other officer went to each of the large white nuclear missiles hanging on the pylons under the wings and pulled the yellow streamer attached to a safety pin from each weapon. Now they could be armed and fired. He threw the two streamers to the assistant crew chief as he climbed up the ladder to his rear seat. This officer was a Radar Operator and referred to both publically and privately as "RO." A RO's duty consisted of targeting any weapons on board and assisting the pilot during the launch of those weapons. He worked the targeting radar aboard each F-89J.

Both pilots were seated in the front seat of each aircraft. They were running through engine start checklists. The crew chiefs stood out front near the large fire extinguisher. Each fanned out the streamers showing each pilot five red streamers and two yellow ones. They received a nod from the pilots and shoved the streamers into their pockets while the pilots began to start the engines. As soon as both engines of each aircraft were churning with a constant roar, the Assistant Crew Chiefs pulled the connector plug from the ground power unit and closed and locked the cover. The crew chief pushed the fire extinguisher out of the way. Both Crew Chiefs stood to the side and gave the pilot a salute. The pilot returned this salute thinking it could easily be the last salute he would ever give.

Each pilot pushed the throttles forward and the ship began to roll out of the hanger. There was no need to get tower clearance from the Elmendorf AFB control tower.

When that Klaxon horn sounded the tower operators immediately cleared the runways and skies of any aircraft that might have been in the way of the two scrambling jets. Security made a high speed run down the length of the runway clearing out any obstacles that might be there. The doors to the fire station house were opened and the firefighters were dressed and standing by in the event they were needed. When complete, the tower was notified the runway was clear and the tower shined a green light towards the alert hangers.

It had been less than five minutes since the horn sounded. Both fighters rolled from the hangers onto the active runway. On seeing the green light from the tower, neither slowed and went "Balls to the wall." The throttles were fully forward the afterburners lit and both were airborne using only 3500 feet of runway. Immediately they turned heading west as they gained speed and altitude.

The senior pilot of the two alert interceptor aircraft was known by the call sign "Gambler." Gambler got his nickname and call sign in pilot training from the other trainees and it was due to his favorite pastime of card playing. The pilot of the other aircraft, a newly assigned lieutenant using the call sign "Reno" was on his first alert standby. He got that nickname because he was from Reno Nevada.

Once airborne, and climbing, the lead pilot contacted the radar control center known as Slugger. He followed normal radio protocol saying the call sign of the unit being called first followed by his own call sign. "Slugger, Gambler with one chick airborne."

The weapons controller at Fire Island AF Station in Alaska responded, "Gambler, Slugger, Vector 275 to intercept three Bogies at Angels 15. Echo characteristics indicate Soviet Bears. Intercept and advise."

"Roger that Slugger, 275 at 15,000" the lead pilot transmitted. He then changed channels on the radio and pressed his push to talk button on his control stick. "Copy that Reno?"

Reno said, "Copied, 275 degrees at angels 15 for three probable Bears. Is this the one Gambler?"

"Relax," Gambler transmitted. "This is your first scramble, just take it easy Reno. Don't start any wars. Just follow my lead. I'll get you back. This is just Ivan's way of making sure we don't sleep. They are about 300 mile from us, cut your AB and save the fuel. We have a few minutes." They were now closing at a total of 950 miles per hour.

Reno's voice was still full of tension as he said "Roger Gambler" and he throttled back to cut out the after burner. They were now over the Bering Strait between Russia and Alaska.

Radar operators at Dragnet, Windmill and Slugger as well as all the others who were aware of the activity were either watching their monitors or listening on their radios. As the Soviet and USAF aircraft closed on each other everyone had the same thoughts. *Please don't let this happen.* There was not a word being spoken. Everyone was as tense as a banjo string. There was total silence from everyone who was watching or involved in the situation. Everyone was thinking either, *please don't fire* or *please don't make us fire.* Discipline was the only thing saving the world from probable destruction. One undisciplined trigger finger on either side would cause death to rain into the Bering Sea. The fish might feed but the world would be at war again.

Gambler and Reno continued west. Both of their RO's were intently looking at their on board radar hoping not to see the three Russian Bears they had been sent to intercept. Hoping they had turned around and went home. It was Reno's RO who first saw the three echos on his screen as they were on their western leg of their orbit.

"Contact!" the RO said into his microphone and the other three officers in the F-89J's all had the same responses. All three instantly became tense. Gambler heard the radio traffic and tried calmly to say "Where?" The RO repeated "fifty miles on course 287."

Gambler made his turn and told Reno, "Continue 287 to intercept and follow in trail." Gambler immediately reported the radar contact to the Controller at Slugger.

Now the two American aircraft, flying side by side, one just ahead of the other flew west to intercept. The standing order was to intercept the TU-95's as far from the Alaskan coast as possible and signal them to leave the U.S. Air Defense Identification Zone (ADIZ), which extended over 200 miles from the coast.

At present the Soviet Bears were over the international water of the Bering Sea. The pucker factor was present in every man in every aircraft from the US as well as the USSR. If the intruders did violate US airspace, then the standard operating procedure was to contact the intruders via radio advising they were in violation and would be fired upon unless they immediately turned. Hopefully the Russians would turn before the ultimatum had to be given and certainly after, but, just in case, both of the F-89J pilots had their thumbs on the missile arming switches waiting for some sign of aggression. Both pilots had their finger on the trigger and prayed they would not have to pull it.

Ivan also had the two fighter interceptors on his radar but the Bears continued east. The interceptors continued west. They were now only seconds apart. Ivan did not want to die. Inside of the Bears, the crewmen knew that if the Americans got trigger happy it meant all of them could be vaporized.

Ivan had played chicken with the US and without any shame in his voice ordered the other two in his flight to "Initiate 180 to the right" and they began their turn.

There was a collective sigh given when the three Soviet planes began a turn towards their home. The two Fighters were still closing on the flight, all in international airspace. That was when Gambler called Reno on the radio.

"Reno, did you bring your camera like we talked about? Over"

"Roger Gambler. Why? Over"

"You will want to take a picture of those Bears to show your wife and kids how close we came to war. Be sure you wave and grin at them and have the camera ready so you can take his picture when he gives you the bird. Over"

Gambler then switched to the standard radio hailing frequency, keyed his mike and said "Good Morning from the Air Defense Command Ivan. Do have a pleasant flight home and God Bless the USA"

Reno got his pictures.

The scenes and dialog of Stargazer and Reno are imaginary. However, it is indicative of the thoughts and words of the pilots and personnel who were involved in many, many actual incursions by the Soviets testing American defenses during the cold war.

The Peacekeepers

We, as Americans have to give thanks to the thousands of men and women of the USAF, young and old, Officer and Enlisted who were sent to these isolate and remote sites to protect the way we live. Radar sites by their very nature are in isolated areas. The weather was always one of the enemies along with time. Time, which was 98% boredom accompanied by 2% sheer terror, was the largest detriment to emotional stability for these people. Off duty each had to do something to keep from going stir crazy or becoming alcoholics. On duty each one knew that they might be involved in some manner of the actual start of another war and in all probability the last war ever. This last war would mean the probable death of everyone and everything they loved. But, this was their job and each one did it the best they could, with total dedication to the United States of America.

Each swore to support, protect and defend the Constitution of the United States. They swore to obey the lawful orders of the President and the Officers appointed over them. This is what they did so that every American could continue to live the way we have. This is what they do to this very day. It is their oath. It is their creed. It is their belief. Thank God for people such as these. Citizens of America, sleep well tonight and every night. You can because the armed forces of the United States, Officer and Enlisted, are awake and doing their jobs. They know the best defense is a good offense. So, thank you and good night America. If that sounds like flag waiving, it is. Now ask yourself, isn't that what we should always be doing?

PART I

The Perceived Problem and the Solution, and The Problems with the Solutions

The Peacekeepers

THE PEACEKEEPERS

CHAPTER ONE

The Problem

Some historians say the atomic bomb ended WW II and many others feel that is true. Some of those historians also feel that it was radar that allowed us to win the war. Not everyone will agree to that but if they examine what radar did to find the enemy, what it did to identify the enemy, what it did to intercept the enemy and what it did to allow destruction of the enemy they might change their minds.

A bit of history about radar. It is short for Radio Detection and Ranging. In 1935, with the advent of the airplane being able to fly across the English Channel in such a short time, England no longer considered themselves an island nation. It was felt that another war with Germany was eminent. About that same time English scientists noted that when an airplane passed through a radio signal the signal was weakened. They conducted tests and verified that it was the radio waves bouncing off the airplane which caused the weakening. As additional testing was performed it revealed that by beaming up radio waves and receiving them, after they bounced off the aircraft, the height, distance and direction of flying aircraft could be determined.

In 1939 a series of radar stations were constructed on the east coast of England creating a radar wall separating it from Germany. The stations were manned exclusively by females wearing the uniforms of the Women's Auxiliary Air Force (WAAF). Electronic surveillance by the Germans using a dirigible detected the radar stations and the signals but misinterpreted them. When it was decided by Hitler to invade England, the WAAF personnel detected the approaching aircraft and reported the information to radar headquarters. The WAAF ladies knew these planes were intending on bombing their hometowns, and perhaps their families. Yet they stayed at their posts doing their jobs, setting the standard for their followers.

With the information radar was providing, fighter aircraft were directed to go to intercept points where the approaching German aircraft would be. Because the English had every air approach covered all the time, German leaders estimated that the English had many more fighter aircraft than they actually did. Had they realized radar was the reason they were being so accurately intercepted they would have destroyed these stations much earlier.

Research continued to try to make Radar more effective and reliable at greater ranges. The invention of an electronic gadget called a Gravity Magnetron did just that. Then, because England's Prime Minister, Winston Churchill, knew that the US would soon be entering the war, the Gravity Magnetron was shared with America. This unit which looked like a hockey puck with wires and tubes sticking out advanced our radar research by two years. By the time the Japanese bombed Pearl Harbor on December 7, 1941 we had five radar stations operational on the island of Oahu, each capable of reaching out 150 miles. On that day "which lives in infamy," the Japanese attacking force was detected by a radar unit and the approaching force was reported, however the report was misinterpreted. The Officer of the Day had heard of a flight of US bombers which would be arriving and felt the radar station had detected them, not an invasion force. That day launched us into a war with Japan.

Realizing the importance of radar and its state of infancy, Massachusetts Institute of Technology and the war department merged resources into a group called Radiation Laboratory (Radlab). Radlab was created to cut through the red tape and bureaucracy. Our military leaders knew the war with Japan would initially have to be fought with submarines while we redirected our industries from peace to war production. In the early years of the war our primary effort was in Europe with the Pacific war being a holding action. For this delaying war, Radlab was to design radar for use in the submarine service. It was not long before radar was being effectively used in the Pacific to prevent other Japanese attacks on the US.

With the war in Europe being won, we were able to concentrate on our war, the Pacific war against Japan. We began taking back islands and forcing the Japanese Imperial Army back to their own shores. When we were at the point of having to attack Japan itself, President Harry Truman decided to, rather than lose even one more American Soldier or Marine, use the recently developed Atomic Bomb to bring the Japanese to their knees and force them to surrender.

On August 6, 1945, the first atomic bomb was dropped. Called Little Boy it was triggered by radar when the bomb was 1900 feet above the earth's surface. When it was triggered the bomb was over the city of Hiroshima and devastated the city. There was no surrender so a second atomic bomb called Fat Man, was dropped six days later over Nagasaki also triggered by radar. The Japanese surrendered fortunately for both of us because we did not have a third atomic bomb. The war was over. The world was at peace, but not for long.

The Soviet Union began a series of not so peaceful actions. The isolation of Berlin; their military support of the North Koreans to bring communism to the entire peninsula; support for the North Vietnamese in their efforts to turn South Vietnam into a communist country, and their Premier, Nikita Khrushchev, making threats against us and other countries of the world caused the US government to become nervous.

It was felt that if the USSR was to attack us, it would have to be by air. The shortest distance to us from the USSR was over the North Pole or across the Bering Sea. They could attack from the Pacific or the Atlantic oceans but that was considered improbable. How could we protect ourselves was the question of the day. The answer was radar. An electronic trip wire built surrounding the United States would give us advance warning. We could send up fighter interceptor aircraft to shoot down any attackers.

In 1947, a separate branch of the armed forces was created, the United States Air Force (USAF). Initially three major commands were formed; Air Defense Command (ADC) would detect and protect the US. Strategic Air Command (SAC) would be the retaliatory force if we were attacked. Tactical Air Command (TAC) was a striking force to be used anywhere in the world.

Radar would be placed under the command of ADC along with the fighter interceptor aircraft which would be used to intercept and destroy any attacking force. ADC Command Center Headquarters eventually would be placed inside of Cheyenne Mountain near Colorado Springs Colorado. All the radar sites encircling the US would report to ADC headquarters which would launch interceptor aircraft from bases located throughout the US. If attacked ADC would detect and defend and SAC, which would be headquartered at Omaha Nebraska would immediately launch a retaliatory nuclear attack against the Soviets.

That was the plan. The strategy and technology was good, but only as good as the people watching and maintaining the radar units, maintaining and flying the interceptor aircraft. Where could the newly formed ADC find such men, who would, like the WAAF of England during their time of war, stand at their post and do their job knowing they might be killed. Or, if they did not do their jobs, allow some Soviet bomber through to bomb a place in the US and perhaps put their families at risk.

The Peacekeepers

THE PEACEKEEPERS

CHAPTER TWO

The Solution

The personnel assigned to the radar sites were special. They had been ordered to go there and perform their duty but the way they performed those duties made them even more special. Some sites were isolated and some were not. Some were good and some were bad. Some were so active they had alerts daily and on some they would have welcomed any radar return even if it was bouncing off a flock of birds. Remember the Cold War? Have you ever wondered why we were never attacked by the Soviet Union? Have you ever thought of what it might have been like had we been attacked or what it might have been like if we had lost?

Our leaders did but not for very long. They were certain neither the Union Soviet Socialist Republics (USSR) nor the United States of America would make the first move and attack the other. It would have meant the end of both countries and perhaps the entire planet if either was foolish enough to attack the other. The reason the USSR did not attack us was primarily due to the ability of the United States to detect and retaliate making it national suicide to make the first move. With sufficient early warning, either country would have been able to conduct retaliatory strikes using bombers. Not much would be left of either country after such action. That was the time we were in, the period called the Cold War.

After the end of WW II the USSR was fearful of our atomic bomb. That paranoia made us enemies. Our government leaders felt Joseph Stalin had a vision of world dominance demonstrated by some of his actions. If the USSR had been the ones with the atomic bomb they might have been able to accomplish that vision through intimidation. But we had the weapon that worked so well it ended a war. The USSR set about to develop their own atomic bomb and did so. They tested it and it worked. Now there was equality, but the saber rattling by the Soviets continued. At that point in time we knew we were going to have a problem with the USSR. Their plans made us scared. We knew we had to be prepared to defend ourselves.

Commanders felt that any air attack from the USSR would most likely come over the Bering straits, over Alaska, across Canada, and into the United States. If they were undetected they could then drop their nuclear bombs on major US cities.

Early detection and the threat of retaliation were considered the solution. How to accomplish that was the problem posed to the military.

Radar had been used successfully during WW II. Radar installations would be the tool of early detection and control of our aircraft that would be used to intercept their bombers.

Military commanders created a defensive plan, a relatively simple plan but very costly. Build radar sites in Alaska and along the northern border of our country. Alaskan radar would give the first warning and we would then send up fighter interceptor aircraft to shoot them down. At the same time another command within the newly created USAF named the Strategic Air Command (SAC) would launch bomber aircraft loaded with our nuclear bombs against the USSR. The military assumed some of the attacking Soviet aircraft would get through the first level of defense. If they did, the northern border line of radar would confirm the attack and other fighter aircraft would be launched from our northern bases to intercept those aircraft. Potential targets would be determined and those defensive elements for those cities would be notified. US Army Nike sites stationed around large cities and strategic locations would prepare their ground-to-air missiles to be fired at any enemy bomber which got through. At the same time southern fighter interceptor bases would launch their interceptors. Our defensive plans accounted for the probability that our radar sites would be among the first targets of an attack. This would in effect make us blind. However, surviving radar sites would continue to watch the enemy aircraft and be in communication with our fighter aircraft directing them to targets.

With all or most or some of their bombers destroyed and our bombers enroute to destroy the USSR there would be no winner. As long as the USSR felt we would retaliate they would not dare to launch a first strike. Sounds simple doesn't it?

In 1947 when the original air defense plan was submitted to Congress by the USAF, it asked for funding approval for a total of 374 Aircraft Control and Warning (AC&W) sites and fourteen Control Centers in the continental United States (CONUS). Another thirty-seven AC&W radar sites and four Control Centers were planned for Alaska. Congress found this plan much too expensive and by the time the legislation was approved a year later, the numbers had been cut to seventy-five radar sites and ten Control Centers for the CONUS and ten radar sites and one Control Center for Alaska.

Radar transmitters send out an electronic radio-frequency (RF) pulsed signal. If the signal hits anything, like an airplane, the signal is reflected back to the radar antenna where it is received and displayed. Early search radar provided the direction and distance of targets within approximately 220 miles from the radar sites. Of course radar signals do not penetrate mountains or hills. In addition to distance, another limitation is created when the antenna is tilted upwards so the focal point is aimed above the horizon. This means as the distance from the antenna increases, the altitude at which targets can not be detected also increases, thus the expression "Flying under the radar" came into being. Reflected radar signals are displayed on a radar console using a Cathode Ray Tube (CRT) similar to a television tube, however the program is very boring and there is no dialog. What it displays is a straight-line beginning at the center of the CRT and extending out to the edge. This yellowish line sweeps around in concert with the antenna which is turning in circles. The yellow line is displayed on a field of eerie green. If anything is out there and it is hit by the electronic signal, the signal is reflected back and appears as an echo or "blip" on the radar console. Operators had to be relieved periodically lest they become hypnotized and ineffective.

Our experience with radar installations in the United States was limited. We had some located along the east and west coasts during WWII manned by the US Army Air Corps.

Art Leighton, USAF Retired, and member of the Air Force Radar Site Veterans (AFRSV) provides the following:

In response to the question, where was the first US radar site? I won't say it was the first, but if not, it was close to it. After a Japanese sub shelled a couple of west coast locations, the army decided to install some radar along the coasts of California, and Washington. One is now a Historical Site near Klamath CA. It was composed of three buildings and camouflaged as a farm. The barn was for the Operations antenna and personnel, an outbuilding which was really a power generation building, and the third, the farmhouse was living quarters. Some of the personnel were quartered in the town of Klamath. The site is protected and cared for by the Parks

Department. This was built in 1942 I believe, and can be visited.

After the war ended the radar installations were closed, except for two, both on the west coast. There was one at Arlington, Washington and the other at Half Moon Bay, California, near San Francisco.

The 1947 plan was approved on the scaled down version, and funded. The US Army Corps of Engineers began building the first twenty-four radar sites. The design of these sites would be based on the two model WWII locations. Each radar site would be manned by slightly less than 200 airmen. There would be five or six Quonset huts in the Cantonment area. These buildings would provide for sleeping quarters, food service, medical, transportation, security, supply, and administration. The operations area would consist of a metal building for radar display consoles and one or two platforms for the radar equipment and antennas to be mounted on. There would be several other smaller buildings used for power production, pumps for water, water storage tanks, and entry control to the radar site.

For planning purposes the radar sites would be spaced about 200 miles apart so their radar coverage would overlap. If there were significant terrestrial features such as mountains that would block radar transmissions, then the sites could be even closer. If there were significant gaps, or "dead zones" in the radar coverage, then a shorter range, minimally manned radar known as a "Gap Filler" would be installed to improve the continuity of radar coverage. The reflected gap filler signal would be sent over communications circuits back to a radar display console at the radar site. The objective was to install sufficient radar sites so that an enemy aircraft could not slip through undetected. Because of high hills and mountains the Corp of Engineers looked for the highest places along the line of the proposed radar site locations. This meant the sites would be placed on the highest mountains along that line. Some of these high spots were such that the personnel who would be assigned would have to take busses or trams to the operations area from the area where they slept and were fed. Some Radar sites would be so far removed from civilization they would receive their supplies twice a year by ship.

Each radar site had to be individually designed to accommodate local conditions. Some sites were all at one level. The antennas and the buildings that housed the radar equipment and display consoles, was called the operations area. The area for the support functions, administration, food service, medical, security, sleeping quarters, motor pool and supply, was called the Cantonment area. If the site was authorized family housing quarters they would be separated from the work areas.

During the initial phase of radar site construction there were locations which were considered to be too remote to provide the essential comforts for female airmen. They were considered Isolated and could only be manned by males. These remote sites were best described as being located so far from outside humanity that eventually a moose would look beautiful. Because of this, assignments would be limited to only thirteen months then the airmen would be rotated back to a better assignment where, if married, he could be accompanied by his family. If single, there would be an opportunity for him to meet a female. There were even some sites within the United States that were considered remote and isolated and were also male only. Other sites were on mountains so high that cable cars had to be built to get the dedicated to their jobs.

Some sites had to be built on two or even three different levels. The highest place would be where the operations area was located and the cantonment area at another level, normally at some flat spot below the top and the family quarters on a third level, if they were authorized.

There it was. Sounds simple doesn't it? It was, but it soon became complex due to radar becoming more advanced and capable of doing more. Initially only the direction of approaching aircraft could be determined. That problem was corrected by using a second type of radar. One which sent out the electronic pulse and when it returned could calculate the height or altitude of the object. Now each radar site would require a second platform, one platform for the search radar and another for the height finder radar. It also required additional space inside operations for the console. More personnel for operations and maintenance of the equipment would be needed. That meant more beds, larger dining facilities, more transportation, and you get the picture. So did the Corp of Engineers and they modified their building plans for the sites under construction. They also scheduled going back to the original sites to modify them. But of course that had to be over time.

Steve Weatherly, then a Lieutenant furnished an example of the hurried construction.

At Mt Hebo AFS, Oregon construction and operation began in the mid '50s when the Cold War was heating up and the threat of a manned bomber attack against the United States was very real. Construction was rapid and many of the site buildings were assembled from corrugated metal sheets based on the classic "Guam Hut" from World War II. From their appearance, these Mt Hebo AFS Guam Huts could have actually have come from Guam. Gunmetal gray in color, with vertical walls, a roof with a curved cross section, numerous vents on the roof, and many dents and scrapes. Low cost and availability were clearly features the Air Force must have been looking for in selecting building. But Mt Hebo was not on a beach, at sea level or in the sunny and warm Pacific Ocean. Instead, Mt Hebo AFS was on a mountaintop at 3154 feet above sea level, with intense rainy and winter weather. In the fall winds at hurricane levels were common. Definitely the suitability of Guam Huts for Mt Hebo was not high on the Air Force list of requirements. So what did this mean for those who lived and worked on Mt Hebo?

For over 2 years I lived in one of these Guam Huts and survived. I had to give up my quarters allowance of almost $90 a month for these terrible quarters. They were so bad that even the government assessment of their "fair market" rental value when used for visitors was the minimum of $15 per month. Some of the "endearing" features of the Mt Hebo Guam Huts that made our life unusual and often difficult were as follows.

Insulation for a Guam Hut was an oxymoron. Insulation was not even negligible – it simply was not there. As a consequence we had one oil fired heater for each 15-foot section of a Guam Hut. There was no forced air circulation so that made for hot spots near the heater and cold spots along the walls. This was especially true in the central latrine area of Guam Huts used for living quarters. Everywhere inside there were drafts that got really annoying in the winter and whenever it got rainy or windy. I never left anything near the exterior walls of my room because water would get blown in between the walls and the floor. In the winter when the winds were blowing and the heaters were fired up, the smell of burning fuel oil would drive you out. I also left my door open most of the time in the winter to let fresh air in.

Foundations for the Guam Huts were poor and the floors were very flexible. No heavy furniture or equipment could be used inside unless extra floor supports were provided. This included safes and filing cabinets. When someone walked down the hallway of a hut, you could see the floor "dimple" under his or her weight. In addition, it sounded like a herd of elephants headed your way.

You could also see the Guam Huts sway in heavy winds. These huts were very light weight and many cables were strung from concrete anchors along both sides and over the roof to keep them from blowing away. Seeing hanging ceiling light and pictures on the walls swaying about was the norm. Since the winds could get very strong, the light weight construction of the metal Guam Hut was like being inside a garbage can when rocks were thrown about outside.

In early 1951, during a status report given to Congress, Commanders pledged that all of the initially funded and approved seventy-five radar stations would be complete by July 1951, however, that was not to be, due to problems within the Corps of Engineers. Delays were due to cost over-runs, and the lack of manpower, building materials and spare parts. There was also a strike at the General Electric Radar fabrication plant that was the major cause for delays. These delays set the project completion date back six months, but by early 1952 the United States was protected by a circle of operational radar sites with trained personnel to detect any sneak attacks.

With some sites operational, the USAF found that the first "lashup" radar sites were built without any creature comforts. That made retention of the trained personnel very difficult if not impossible. They ordered that starting in 1952, all future sites would include additional buildings for such things as recreational facilities to house libraries, theaters, hobby shop areas, and pool tables. Some even got bowling alleys. At the earlier completed sites drinking in the barracks had been condoned but resulted in problems such as fights, drunkenness and even alcoholism. The commanders of every site combated this problem by outlawing booze in the barracks and creating or making space to be used as a bar or club. If this meant shrinking the size of supply or the dining area, so be it.

To further complicate matters the military commanders of ADC and of Canada agreed they too should have some protection and notice if an attack was imminent. Canada agreed to allow the US to place thirty-three AC & W sites along a line known as the Pinetree Line. The Pinetree line was located across the width of Canada just north of the US's northern border. The Pinetree line was manned by USAF military personnel with some Canadian civilian employees. These sites could track the attacking aircraft and possibly determine their targets which of course were thought to be the military airfields, posts, and yards, with secondary targets of major cities.

Then in 1957 an additional ninety unmanned stations were built 300 miles north of the Pinetree line. This line was known initially as the McGill Fence and later as the mid-Canada line. The next year fifty eight Distant Early Warning sites (DEW line) which would give about a three hour advance warning of any attack from the USSR were built in North America looking over the polar cap.

All of these manned and unmanned sites were generating air defense information. This presented significant challenge to the Command and Control functions of the USAF and the Canadian Air Force. That was remedied in May of 1958 with the development of the North American Air Defense Command, shortened to NORAD. NORAD would be a combined command led by an USAF General Officer and the vice-commander would be Canadian. NORAD headquarters would be in Cheyenne Mountain, Colorado. They would have communications with all the radar sites and bases of the ADC as well the Canadian Air Force and the authority to launch interceptor aircraft.

By this time, most radar sites had been modified to include search and height finder systems. Progress in the development of the radar systems themselves allowed these systems to include transmitters, receivers, and display subsystems. The transmitter and receiver subsystems were designed to share an antenna. Search radar antennas typically revolved at six revolutions per minute. Height finder radar antennas nod up and down and can be quickly rotated to different azimuths. Different cathode ray tube (CRT) display units were used for search and height finder radars. The search radar display (Plan Position Indicator or PPI) is typically circular, has a rotating electronic sweep, and detected targets appear as small blips (also known as paints). The electronic sweep origin is at the center of the circular display and is synchronized to the movement of the rotating antenna. The blips appear at the azimuth and range of the detected targets. The blips appear as small points of light on the display that persist after the sweep passes by. The blips are refreshed (repainted) with each pass of the sweep and over time this shows the direction and speed of the target.

Height finder displays (Range Height Indicator, or RHI) were typically rectangular, and the electronic sweeps moves up and down and was synchronized to the movement of the height finder antenna. The electronic sweep origin was usually a point on the lower right side of the rectangular display. Targets detected by height finder radar were displayed as vertical blips, much longer and broader than for the search radar. The mid-point of the vertical height blip indicated the altitude of the detected target. The azimuth of the detected target was the same as the position of the antenna and was available to the operator as a separate readout. Separate radar display consoles were available to assist controllers at radar sites. These specialized displays helped the controllers detect targets and guide aircraft to intercept points. Controller scopes were available at radar sites where manual operations were conducted.

AC & W Radar Site Operations

During the early years, ADC manned these radar sites with people assigned to an Aircraft Control and Warning (AC&W) Squadron. These AC&W Squadrons included personnel for the maintenance and operation of radar, and radio equipment. There were also controllers who manually provided commands by radio to interceptor aircraft. ADC was responsible for these aircraft Interceptor Squadrons assigned to bases.

At some ADC bases the Air Force stored an air launched nuclear missile specifically designed for their use and known as the GENIE (MB-1) for the F-89J Scorpion interceptor and later the F-102. At least two aircraft were designated as "Alert Aircraft" and had Genie's hung on pylons or in the weapons bays while in that status. Practice runs by the alert crews were conducted and generally showed that from notification to the alert aircraft being ready to taxi out of the alert hangers was less than five minutes.

While not co-located with most AC&W radar sites, the US Army Nike Air Defense System was integrated into the overall NORAD system.

At the radar sites, in the operations area, sitting in front of radar scopes watching a ray of yellow light go around against the green backdrop were the Radar Operators. Other Radar Operators, when not on the scopes, stood at a large transparent sheet of Plexiglas, wearing head phones and drawing symbols and writing backwards. They had to write backwards because they were behind the screen and others who needed to read what they were writing were in front. The people in front also wore headsets or had hot line telephones in front of them. If they picked up their phone it would automatically ring at another location where others waited.

The Radar Technicians sat at their consoles hoping that the yellow ray would not bounce off any routine target. They knew there were certain routine flights of airplanes they would receive an echo from and those were okay. They could distinguish these from the transponders that squawked a radio signal Identified Friend from Foe (IFF). It was the unexpected returns from specific areas that would get their adrenaline rushing if they did not get their wish. Either no IFF signal or the wrong signal for the aircraft's filed flight plan would cause instant concern to put it mildly.

Behind the radar operators was the crew chief. He might be a Non Commissioned Officer (NCO) or it might be a ranking airman, also a radar operator. His job was to help the Operators interpret the returns that displayed on their radar sets and to order relief for the operators when they began to get bleary-eyed.

Behind the crew chief was an officer, most often a lieutenant or Captain and sometime a major, called the Weapons Assignment Officer (WAO). He had the authority and responsibility to alert fighter squadrons of any unknown targets. Then once airborne the fighter pilots would be instructed by another officer called the Weapons Controllers as to where go to intercept and identify the unknown. Depending upon manning the Weapons controller position might also be combined into the WAO position. During times of increased security, if an unknown aircraft refused to identify itself or displayed some other aggressiveness, then the pilots of the interceptor aircraft might be given the authority to destroy the intruder.

Another officer called the AJO or Anti- Jamming Officer might also be nearby in the event there were attempts to jam the radar, or the communications between Radar Operations and the interceptor aircraft.

Working in the area, as needed, were radar maintenance personnel. Their job was to make certain the radar systems and consoles were operating and if not fix them so they would.

The Communications Center was often in the building. Affectionately nicknamed Ditty Boppers due to the dits and dahs of Morse code which they all had to learn, they operated the teletypes and decoding machines essential for non verbal communications, and manual phone systems.

Within a separate building would be power production personnel. They operated and maintained the power generators that were always running or were of the uninterrupted power supply (UPS) type that would automatically come on during loss of civilian power.

In another area would be the radio people nick- named "Gators." GATR stands for Ground to Air Transmit and Receive. The GATR area had to be located away from the operations area because of the interference the radar would cause with the radio signals.

In the support area which might be physically separated from the Operations area would be the sleeping quarters, the motor pool, the headquarters building, security, medical, club and most important the mess hall or dining hall. If they had a recreational building it generally would be in the support area as well.

The length of tour for the personnel assigned to radar sites varied depending upon the classification of the site. Many of the AC & W sites located in Canada and Alaska were considered isolated remote tours and understandable especially when for six of those months there is snow on the ground and the weather would hover around zero. Anything above freezing would be considered a summer day. Temperatures of fifty degrees below zero were common. So common, airmen were told to never run outdoors in the winter lest they get out of breath and start gasping. The extreme conditions would cause a person's lungs to freeze. The shortest tour was thirteen months at isolated and remote radar sites but that could be a very long time.

AC&W Squadrons usually had about 150 people assigned. Because of their locations, there was little activity for entertainment. Stuck at some assignments for months or years, many people turned to alcohol to get them through. Some became jokesters, pulling pranks on each other. Some became sportsmen learning to hunt and fish, and to ski and snowshoe. Some learned to play cards, pinochle being one of the favorite games because a game could go on for days, round the clock. That is how they spent their off duty time. On duty they were a group of dedicated professionals playing the serious game of keeping the peace and securing the American people and their way of life.

The Peacekeepers

THE PEACEKEEPERS

CHAPTER THREE

What would they look like?

Generally speaking there were three types of AC&W sites. The first type was the remote radar site. Most of these were located outside the continental US (CONUS) in places where the Air Force could not accommodate civilian dependents. The wives and children of the personnel assigned to a remote site could not accompany their husband and father to the station and live there. This meant the airmen would be assigned to the site while his family stayed elsewhere. In the case of most married lower grades this meant with the wife and kids went to his or her family home and lived with them for the duration of assignment. Assignments at remote sites were limited by not allowing back to back remote assignments unless specifically waived by the airman. Airmen at remote sites were on a thirteen month tour and not permitted to go home on routine leave for the duration of the assignment, emergencies excepted.

A second type was the isolated radar site. These sites were located away from big towns and cities, away from population centers, in other words, rural in nature. Bringing a family to some of these isolated locations was discouraged. This was due to the lack of available housing or due to the cost of supporting a family in some of these isolated areas. Isolated assignments could be for thirteen months if unaccompanied or two to three years if the airman elected to bring his family over the objections of the USAF. On isolated assignments, an airman could take routine leave one time during his tour. They could also take a weekend pass with specific permission.

The third type of radar site location was on an Air Force Base, or near a large population center where all the conventional support services were available. Services like shops, food stores, service stations and housing. There was no limit to how long an airman might be assigned to these sites. Assigned personnel were allowed routine leaves, vacations and passes. They normally worked only eight hours a day and had days off. That was something airmen assigned to remote or isolated sites did not generally have, days off. They worked every day, often more that eight hours each day and only rarely had two days off.

Initially, only male airmen were assigned to the first two types of stations. Female airmen were only assigned to the third type.

Once operational some sites noted gaps due to natural geologic formations, e.g. hills, mountains, plateaus, and buttes along the radar signal path. Sites called Gap Fillers were used to plug those holes. Gap Filler sites were unmanned but had to have some maintenance performed on them. They were connected to the radar site by telephone lines and were monitored on a separate console at the primary radar site.

An important addition to the air defense system was the acquisition of the EC-121, a four engine turboprop long-range surveillance aircraft. In addition to providing extended radar surveillance off the East and West coasts, these airborne early warning aircraft also flew off of Iceland. These airborne radar sites carried between eleven and twenty-five surveillance and intercept control personnel in addition to a six man flight crew. They flew specific routes, mostly over water, reporting surveillance tracks by radio and later by data link to Control Centers.

Depending upon the physical location of each radar site it might be surrounded by a fence. Some locations were so remote or they sat on the very top of a mountain with only one way up or down, they were deemed not to need any additional security. One site in Alaska was on a mountain top accessible only by a cable car. The operations areas were almost always fenced with only one entry point for personnel and, if accessible by road, a gate for vehicles. Depending upon the site layout, these operations gates were manned by armed security policemen or by operations personnel. Often these gates were unmanned, but access was controlled by an electronic lock which would require the person desiring entry to press a series of numbered keys to open the gate. Misdialing would cause an operator with a weapon to exit the operations area and check the identity of the person wanting access.

There was a peculiar action noted during the early years on the sites built with the operations area and radar towers on the same level as the cantonment area. The Radar Signals would cause Fluorescent lights to light up every time the radar beam would pass over the area because of the electro-magnetic energy put out by the radar. All of the fluorescent lights had to be changed to incandescent bulbs. This energy also became a concern for personnel safety. This energy might also detonate explosives. Accordingly, the radar transmitters were prevented from transmitting on azimuths where explosives were stored. There did not appear to be any interest in protecting the personnel from any of these signals and they were felt to be non-dangerous however, by the 1950's it was well known that radar transmitters needed shielding to protect the personnel as well, and the personnel were provided with radiation detection badges to keep track of their exposure.

During the early years, the construction of the sites was fast. Tents were used as temporary quarters for the assigned airmen. Wood framed buildings covered with plywood became headquarters, supply rooms and kitchens. The luckier ones had Quonset huts, wood frames covered in corrugated steel and shaped like half a vertical circle. These withstood the weather better than tents.

The towers for the radar antenna and for the operations building were what everyone was interested in erecting. They were steel beam frames, welded and riveted and covered with corrugated steel sheeting. Inside on a raised floor was the working area for operations. This is where the men watched the radar scopes waiting for a blip or return to appear. In front of them was a large sheet or sheets of Plexi-glass. With the outline of their area of responsibility painted on. Airmen stood behind the glass and received aircraft location information from the operators through headsets. That information was transcribed onto the plexi-glass, writing backwards so the officers and senior NCO on the raised dais in front of the plexi-glass could see what was being reported. These buildings were blistering hot in the summer sun and freezing cold in the winter. Initially in the winter, the airmen wore parkas while working. Eventually heating units were installed. The temporary building and offices also got fuel oil stoves that were placed in the center of the un-insulated buildings. In the barrack or office buildings, having a bunk or desk too close to the heater meant you would burn up. However, if your work or sleep area was more than ten feet away, you froze.

Within the operations area or close by in a separate building would be the power generators that were always running, or the type that would come on during loss of commercial power. These large diesel generators had only one priority customer, the operations areas.

Communications personnel were often collocated in the operations building. These men operated the teletypes and encryption machines essential for administrative and command and control communications using telephone lines to Command Centers, interceptor bases, and supporting units. Communications maintenance personnel maintained the teletypes, decoding machines, telephone switches and phones, technical controls, mainframes, and telephones lines on the radar site.

The information collected from the individual radar sites usually had to be forwarded by telephone to Control Centers. Most common for these Control Centers were located on a radar site. They not only had their own information but they also received information from other sites within their area of responsibility. A few were located separately near large metropolitan areas. Inside those Control Centers was a larger Plexiglas screen and dais for some additional command personnel.

Typical of a remote site was Tin City Alaska. Manned by only men, Tin City was a two level site. The operations area was located on a mountaintop and the support functions were at a lower level. This was also one of the sites that the operations area was so remote they used a tram to get to and from the two areas. Every eight hours a new crew would relieve the old crew. There were times the weather was so inclement that rather than use the tram, the relieving crew was trucked to and from the operations area, a feat which might take as long as an hour each way. About 75% of the personnel assigned to sites would be operations personnel. The other 25% would be support personnel.

Support personnel worked in the cantonment area which was physically separated from the operations area. Medics usually had just a room in a building. Most sites only had an enlisted medic who could fix minor cuts or scrapes. Serious injuries had to be evacuated to a nearby military installation which had a hospital, or they were treated in the nearest civilian facility. Medics within the military are universally known as "Doc" provided he was respected.

Air Policemen, referred to as "Apes" or sky cops took care of entry control to either the site or to the operations area. They were responsible for preparing a site defense plan, inspecting weapons, and training other airmen in the defense of the site. They would work out of a small shack at the entry control point to the location.

Cooks, referred to as "Cookie" had one of the most important jobs on a radar site. Feeding the troops and keeping them happy. If a site was plagued with a bad cook it could affect the morale of the airmen. The Dining Hall more commonly called the mess hall was, in the early days, a tent. Hot water needed for cleaning, essential in any mess hall was heated in clean garbage cans with kerosene powered water heaters. Initially, the airmen lived on combat rations, commonly known as C-Rations. An excellent cook could make C-Rations barely palatable.

Administration, supply, transportation, civil engineering, security and other personnel supporting the radar operation were also located in the cantonment area.

There were also a number of commercial services and unofficial functions that supported the site. Because of the site locations, commercial television or radio broadcasts were not always available for entertainment. Often site personnel would devise their own radio stations complete with volunteer disc jockeys. Speakers were hard wired into every room and office on the site and the volunteers would schedule round the clock entertainment. A Military Affiliate Radio Station (MARS) was operated by volunteers.

Movies might be received weekly or bi-weekly and shown in a theater that doubled as a chapel for Sunday services if there was a minister or priest, nicknamed Padre, on the site or visiting the site. It would also double as the meeting room for the monthly mandatory commander calls or other briefings as required. When it was a theater, the movie projectionist was an airman who was paid as much as forty cents per night if he could get at least ten people to see the movie at ten cents per person.

Movies were usually 16mm films that had been circulated in the civilian theaters for some time before they were shown on the sites. The movies were shipped from site to site with a new one arriving about every week. Some of the more civilized locations got two different movies every week. Keeping the projector and sound system isolated from the every 12 second audible search radar buzz was necessary and a challenge. There were plenty of skilled volunteers to provide a top notch RF isolation system for the projector.

An airman who learned to cut hair was indeed essential and was paid whatever he charged for a haircut. If he was pretty good he might charge as much as a quarter. If he was just learning, a dime was what he could charge, but every month he was guaranteed to have every man in his chair at least one time.

In one of the Quonset huts there would be laundry available to the barracks occupants. However at most every radar site, there would be an airman assigned to run the laundry. He would wash and dry the sheets and pillowcases as his Air Force duty. As a part time job he would also wash, dry and iron uniforms for a price. There would always be one man who would earn additional income by operating a branch of a Base Exchange where he would sell cigarettes, hygiene items, candy and uniform items. He would also special order civilian clothing or sports equipment like fishing lures and rods and reels.

It was also essential to have a good relationship with the local community. Most of the time many of those assigned to the radar site and in some cases families lived in these communities. Churches, stores, schools, hospitals, doctors, restaurants, recreation centers, service stations, fire stations and police were available. Usually these communities also had radio, TV, and newspapers. Social gathering at clubs, dances, and sporting events offered a way to "get off the hill and unwind." These communities also had personnel that were hired to fill radar site positions for civil engineering and power production.

Within the CONUS, local telecommunications companies were essential to meeting the external communications requirements of the radar sites at a reasonable cost. The local Telephone Company (TELCO) often provided access to local and long distance commercial phone service. The local TELCO provided the long haul trunks to Direction Centers, support bases, and connectivity to the military AUTOVON and SAGE BUIC AUTOVON networks. The local dial central office (DCO) at some sites was owned and maintained by the local TELCO. Air Force inside and outside plant personnel were not assigned to a radar site if their jobs could be provided by the local TELCO. If a good relationship was established with the local TELCO they contributed to the Air Force site Base Wire Communications Plan (BWCP). This was a big help to the radar site communications officer.

The presence of a radar site near a small community was usually a big plus. The telecommunications needs of the radar site meant more high skilled local jobs, and brought current technology and services to the area. This included dial and long distance service. During the period from the end of WWII until the later 70's many of these local TELCO's were small, family owned operations with an antiquated infrastructure and limited connectivity into the much, much larger telecommunications corporations with a large regional or nationwide network. Today most of these small companies have been bought out and absorbed into their larger cousins

In early 1951, during a status report given to Congress, Commanders pledged that all of the initially funded and approved seventy-five radar stations would be complete by July 1951, however, that was not to be, due to problems within the Corps of Engineers. Delays were due to cost over-runs, and the lack of manpower, building materials and spare parts. There was also a strike at the General Electric Radar fabrication plant which was the major cause for delays. These delays set the project completion date back six months, but by early 1952 the United States was protected by a circle of operational radar sites with trained personnel to detect any sneak attacks.

Millions of dollars were spent on building the radar sites, control facilities; communications, logistics support bases, and the fighter interceptor bases. The technical equipment installed on each one was essential to the mission and also very expensive. All of it would be worthless without the most essential commodity, the personnel.

There have always been many questions as to how Air Force volunteers were selected and trained for duty at remote and isolated Radar Sites where there were few comforts. In order for the radar sites to accomplish their mission, the personnel with their various skills were essential.

Eighteen-year olds registered for the draft. If drafted into the Army they would have to serve two years in the infantry unless they had some specific training, talent or occupation. If they volunteered for the Army they would have to serve three years but they might be offered their choice of schools and military occupations. All draftees went into the Army. Volunteers had their choice of the Navy, Marines or Air Force but had to enlist for four years.

It happened at recruiting stations throughout the United States. Recruiters had a conversation with young men wanting to do their share in the protection of the US not to mention avoid the draft into the Army. Air Force recruiters did not say, "How would you like to be stationed on a mountain top with 150 other GI's, all males often fifty miles from the nearest girl? Your job will be to stare at a television set, but there will be no story, and there will be no dialog. You will watch a line going around. You will learn to write backwards. You will be forced to eat canned food for months on end. You will be in some of the coldest places on earth. You will be there for at least a year." The truth would have scared the recruits straight into the arms of other service recruiters. Rather, the recruiter said, "we have a job for you we call radar operations. It is all secret work and you will be working with some of the best minds in the Air Force. Your tools will be some of the newest electronic gear in the world and you will be part of it. Sign here. And they did.

Many of these eighteen-year olds wanted to obtain training in certain career fields. The Navy and the Air Force offered the best choices for these training schools. Both of these services required a high school diploma to be able to enlist. Before entering any military service, enlisted applicants had to take certain tests. These tests were not the pass-fail type; rather, they were measurement tests. They were designed to determine if a volunteer applicant had any mechanical, administrative, or electronic aptitudes, or if they fit into the general category. Then the specific service recruiter had to convince the applicant what was best for him. There were some applicants who volunteered predicated upon a specific job and generally they were assigned to that field if they were otherwise qualified for it. If there was any hesitation on the part of the applicant, the recruiter would simply tell him, once in service, they might be able to transfer into some field they really wanted. That did occur but very seldom. The needs of the service always came first and the recruiters funneled applicants into the career fields which needed filling.

For the Air Force volunteers, if they scored sufficiently high on the electronics portion of the tests, the recruiter convinced them what career field they should select based on the needs of the Air Force. This was fairly easy to do. They merely had to tell the applicant a particular field was full even if it was not. During the 1950s and 1960's, the Air Force wanted and needed Aircraft Control and Warning site personnel. At its peak they needed over 140,000 radar operators and maintenance personnel. As a result many volunteers were guided to those career fields and away from the fields directly associated with airplanes and aircraft maintenance which most of the applicants really wanted.

Once convinced, applicants went to an induction center for a physical. While it may have seemed to the applicants that if they walked, talked and had all their limbs they were sworn in, these physicals did more than that. Their teeth were checked and uncorrectable problems determined. Their feet were checked and anyone with flat feet rejected. Applicants who were colorblind could not go into electronic maintenance because of the color- coding of wires, and components such as resistors and capacitors, they would have to work with. Physical deformities were cause for a person to fail. Problems with internal organs were checked for and if found might cause an applicant to fail. Applicants who did not pass, and there were a small percentage, were sent back home with a draft classification change. They would be safe from the draft except in the event of a national emergency.

51

After this mental and physical screening process, applicants would be sworn in to active military duty as Airmen. This oath required them to obey the orders of the officers appointed over them and to support and defend the Constitution of the United States against all enemies, foreign and domestic. Then they were sent for Basic Military Training. Most would be sent to Lackland Air Force Base (LAFB) in San Antonio, TX. Basic Military Training (BMT) varied in length from four to eight weeks with some as long as thirteen weeks. Until the late 1950s some were also sent to Sampson AFB, near Geneva, NY. It would take that long to convert these civilians with little or no military discipline into Airmen. They would have to know how to maintain a military bearing, keep their living area and clothes clean, make beds the military way, maintain closets in a structured way, properly wear their uniforms, march, salute, and say "Sir" a lot. They had to get into physical shape and that meant running everywhere and doing pushups for real or "perceived" infractions when ordered by training instructors. All Airman Recruits had to know how to shoot a rifle, how to eat, when to eat, what to eat, stand at attention, and what seemed to be the most important thing; to hurry up and wait.

After learning all of the required military tradition and acknowledging and performing to Air Force disciplinary standards, the enlisted recruits selected for the electronics career fields were sent to Keesler Technical Training Center (KTTC), on Keesler Air Force Base outside of Biloxi, Mississippi.

In order to speed up BMT and technical training, some trainees were sent to Keesler AFB before they had completed BMT at Lackland AFB. These trainees then completed the rest of BMT while also starting training in their electronics career field.

Officer personnel enter the service in one of three ways; either through a service academy such as West Point or Annapolis and later through the Air Force Academy, through college by participation in Air Force Reserve Officer Training Corps, or through Officers Training School (OTS).

Personnel becoming officers through OTS attended a school which lasted three months and on graduation received their commissions as a second lieutenant. These officers would be referred to as "Ninety-day wonders." Some of these ninety-day wonders turned out to be the best supervisors because many had been enlisted personnel before going to OTS.

To be a pilot, the airman had to be an officer, but not all Air Force officers would be pilots. For the most part, they became managers and supervisors of the airmen who do the work.

Some officers would be selected for radar site duty and attend advanced technical training depending upon their career fields. Electronics Officers were trained at Keesler AFB (KTTC). The basic Weapons Controllers Course was taught at Tyndall AFB in Florida near Panama city. There were also other special courses for Controllers taught at KTTC.

Ground Electronics Officer training was a year long school and consisted of advanced mathematics, electrical fundamentals, alternating and direct current, amplifiers, oscillators, vacuum tubes, transmitters and receivers. They watched many US Navy training films regarding these topics. Once they had the basics down pat, they were trained on the various types of search radar and height finder radar transmitters and receivers. They had to know the radar display consoles inside and out. There was a lot of classroom work with circuit diagrams and technical orders to identify and resolve equipment problems. They had to learn the preventive maintenance schedules and instructions for all the equipment and how to use the testing equipment. Written tests evaluated their progress. They learned the basics of electronic warfare and aircraft intercept procedures and systems. Others assigned those duties would be taught the intimate details of those functions.

For the first week of assignment to KTTC, the new Airmen Basics or Airmen Third Class were assigned to the dining hall on Kitchen Police (KP). There were four dining halls which served around 13,000 students who were taking a hundred various types of training plus the instructors and the permanent party assigned to KTTC. Some training classes were held from 0600 until 2400 which meant the dining hall had to be open from at least 0500 to 0100 seven days per week. It became a constant job of cooking, cleaning, feeding, cleaning and start all over again.

Initially, the training was done in three shifts by instructors who were civilians or in some cases senior NCO's. The morning class began at 0600 and went until noon and the afternoon shift was from noon to 1800, with the night shift from 1800 to 2400. Their training covered the same topics as the officer training but not in the same depth. Enlisted radar operators, scope dopes, had to know how radar operated and how to interpret the displays, report the displays properly and how to write backwards. Enlisted radar maintenance personnel had to know about circuit diagrams and the preventative maintenance schedules and the use of electronic test equipment. With less stringent schedules, enlisted Radar Operations training varied in length and was between twenty-four and thirty-six weeks.

Ground Electronics Officer training was a year-long school and consisted of advanced mathematics, electrical fundamentals, alternating and direct current, amplifiers, oscillators, vacuum tubes, transmitters and receivers. They watched many US Navy training films regarding these topics. Once they had the basics down pat, they were trained on the various types of search radar and height finder radar transmitters and receivers. They had to know the radar display consoles inside and out. There was a lot of classroom work with circuit diagrams and technical orders to identify and resolve equipment problems. They had to learn the preventive maintenance schedules and instructions for all the equipment and how to use the testing equipment. Written test evaluated their progress. They learned the basics of electronic warfare and aircraft intercept procedures and systems. Others assigned those duties would be taught the intimate details of those functions.

Officers could come and go at will after their training classes; however, the enlisted men were marched in formation to and from the classrooms. Officer quarters were rarely checked while enlisted quarters were frequently inspected. Officers could drive their own vehicles and did not need passes or permission to leave the base. The enlisted personnel were not allowed passes to go off base until they reached a certain point in their training, then they had to rely on either their vehicles which had been parked or commercial transportation. But probably the biggest difference in the minds of the enlisted people was the fact that officers did not have to pull Kitchen Police or perform other squadron duties such "police calls" which were to clean the area around their barracks. The common instruction given at police calls was "If it is not green pick it up. If it's too big to pick up, paint it green. If you can't paint it, salute it and move on."

After the required amount of training time and successful examinations, enlisted trainees would graduate and be awarded their Air Force Specialty Code (AFSC) as trained airman with a skill level equivalent to an apprentice. For example, an airman trained to be a heavy ground radar repairman had an enlisted man's AFSC of 30332. Officer trainees upon graduation were awarded an entry level AFSC. For example, a ground electronics officer had an AFSC of 3041. They were now eager and ready to protect the US. With orders in hand they reported for their duty assignments at an AC & W Squadron.

In many cases, upon arrival at their assigned radar site, the personnel were often disappointed to discover their new home lacked certain amenities. These and other hardships caused many airmen to leave the Air Force after their first enlistment.

Because of the site locations, commercial television or radio broadcasts were not always available for entertainment. Often site personnel would devise their own radio stations complete with volunteer disc jockeys. Speakers were hard wired into every room and office on the site and the volunteers would schedule round the clock entertainment. Movies might be received weekly or bi-weekly and shown in a theater which doubled as a chapel for Sunday services if there was a minister or priest, nicknamed Padre, on the site or visiting the site. It would also double as the meeting room for the monthly mandatory commander calls or other briefings as required. When it was a theater, the movie projectionist was an airman who was paid as much as .40 cents per night if he could get at least ten people to see the movie at .10 cents per person. An airman who learned to cut hair was indeed essential and was paid whatever he charged for a haircut. If he was pretty good he might charge as much as .25 cents. If he was just learning, .10 cents but every month he was guaranteed to have every man in his chair at least one time. In one of the Quonset huts there would be laundry available to the barracks occupants. However at most every radar site, there would be an airman assigned to run the laundry. He would wash and dry the sheets and pillowcases as his Air Force duty. As a part time job he would also wash, dry and iron uniforms for a price. There would always be one man who would earn additional income by operating a branch of a Base Exchange where he would sell cigarettes, hygiene items, candy and uniform items. He would also special order civilian clothing or sports equipment like fishing lures and rods and reels.

THE PEACEKEEPERS

CHAPTER FOUR

Nothing is more permanent than change

We were feeling fairly well protected from enemy bombers in the late 50s. We still had gaps in our perimeter defense system. But what if the enemy did not approach as the commanders thought, let's say they flew over Europe or Asia and attacked from the east, west or south coast. We would be vulnerable. The solution was to build flying radar sites called AWACS, Airborne Warning and Control System aircraft. They would plug several of the gaps, but there were still holes in our circle of protection. Perhaps the Soviet Union would not be our only enemy. To correct this, other radar sites within our own borders especially the east, south and west borders were constructed. There were some areas on the east and south coast where we just could not have a radar site on shore with adequate radar coverage. In response, Texas Towers in the Gulf of Mexico and along the East Coast were built on the ocean. Texas Towers were quite similar to ocean oil rig platforms. The sites were placed on stilts and the radar systems built on top of the platforms thus extending radar coverage and adding early warning time. Now we were protected.

Everything changed on October 4th, 1957. Prior to that date our perceived threat was from bombers. On that day the world heard the electronic noise being emitted from a Soviet satellite called Sputnik. The US had been working on missiles capable of sending satellites into space. When the beeping was heard, the US knew the Soviets were ahead of us. If they could send a missile into space with a satellite, they could send a missile aimed at us with a warhead and we would have no warning.

Hours and hours were spent trying to learn what the beeps meant. Was it codes? If so what were the coded transmissions regarding? In the end, we learned that the beeps were only noise. It was not transmitting any information back to earth.

It was too late. Our paranoia kicked in. If they could send up a satellite with a missile, then they could use a missile to neutralize our retaliatory weapons before we could use them. Again, our strategy had to play catch up. Another detection system, the Ballistic Missile Early Warning System (BMEWS) was developed and placed in Alaska and North America at three new sites. It would alert us to any missile launch from the Soviet Union and aimed at Canada or the United States. We had developed our own Intercontinental Ballistic Missiles (ICBM's) capable of being used as retaliatory weapons. These were located in underground silos throughout the US. With BMEWS giving us advance warnings, we would have time to launch our own missiles against the Soviet Union and time to send the SAC bombers to their targets. We would still require radar sites in the event the USSR launched a bomber attack and still require our bases and fighter interceptor aircraft. Again we were at a stalemate with the "Red Menace."

No business or industry, military or civilian activity can exist without undergoing many routine transitions. In the case of Radar sites, transitions were due to the changes in military systems, threats as we came to know them, the implementation of new technologies, public opinion, and the need to retain and motivate personnel came into play.

In 1959, a second phase of radar defenses was designed, developed and initiated. This was based upon a computer-controlled network of radars, operations centers, and interceptor aircraft This was the Semi-Automatic Ground Environment (SAGE) that was operational until 1983. In the SAGE system, AC & W Squadrons would be renamed Radar Squadrons and most no longer would have assigned controllers. The controllers would be assigned to command centers known as Direction Centers, or Combat Centers (DC's or CC's). The Lashup, permanent and mobile radar sites were upgraded with newer radar, radio, and data communications. The older site designations such as P, and M became Z sites. For example, P-49 Watertown AFS, NY became Z-49 and M-100 Mt Hebo AFS, OR became Z-100. Additional radar sites were constructed and some of the old sites closed. A Backup Intercept Control (BUIC) capability was added to some radar sites to make SAGE more robust and survivable. Interceptor aircraft could now be automatically controlled to intercept points. ADC added missiles such as the improved aircraft launched GENIE (MB-1), a nuclear missile and the ground launched BOMARC (CIM-10B) missile to the weapons inventory. The nuclear-tipped supersonic BOMARC's were guided by controllers in the SAGE blockhouses and could engage aircraft out to the limit of ground radar coverage. In addition, the US Army upgraded the Nike series of missiles and added the Hawk missile to their weapons inventory.

The SAGE radar sites used similar personnel except there were no weapons controllers. There was no need for airmen to have to write backward on plotting boards as they were no longer needed. That information would be sent to the DC's electronically. Sites no longer had command center functions. Intercepts normally displayed on the old radar consoles would now be automatically displayed on the radar consoles at the SAGE DC's and CC's. Commands to intercepting pilots would come from there and much of that was computerized. It was still a semi-automatic system because of the height finders located at the sites. Operators would have to manually provide the SAGE centers with the current height of a designated target and because officer controllers were required to make tactical decisions.

In the spring of 1958 **Shaun Finn** was at the 762nd AC&W Squadron, North Truro, Mass.

They were "experimenting" with the transition to SAGE operations. I remember the search radar target data being automatically sent via communication circuits to the SAGE computer at the Direction Center. This was the "automatic" part of SAGE at a radar site and was always active.

I also remember that the operation of the range height indicator (RHI) became more demanding. With SAGE, the antenna azimuth direction was automatically controlled by the computer at the DC. The "reported altitude" would then be displayed on the RHI by a horizontal cursor bisecting the target paint". The operator would then manually make any needed adjustments to the computer generated RHI cursor location based on the actual, real time target altitude and azimuth "paint." More often than not, the RHI operator also needed to slew the antenna azimuth slightly to one side or the other from the SAGE computer generated position as part of the task to get a better updated "paint" for target altitude measurement. When all the radar azimuth and RHI cursor adjustment were completed, the operator pushed the send button to get the updated target data sent to the SAGE computer at the Direction Center. This was the "semi-automatic" or manual part of SAGE at a radar site. Sounds complicated but was not. In actual use and took only seconds to complete. It did, however, require an attentive operator sitting in front of the RHI. In quiet times there were often long period with no height requests from the DC.

During the transition from AC&W to SAGE Radar Squadrons there were many radar site equipment changes that enhanced the radar and operational capabilities. Among these was the replacement of some earlier model radars with newer systems. There were the new frequency diversity search radars such as the FPS-24, 27, and 35 that provided a wider spectrum of transmitter frequencies that enemy forces would have to deal with in their radar order of battle. The FPS-26A provided a new height finder radar with higher power and a higher frequency than older height finder equipment. Some FPS-26A radars were subsequently modified into the FSS-7 for missile detection and operated by a separate Missile Warning Squadron detachment at the radar site. New radar towers were constructed to replace older designs that could not accommodate the increased size and weight of these new radar systems. Power plants were upgraded or replaced to meet the higher electrical demands. Communications systems were improved to provide for the connectivity need for SAGE operations between the DC's and the radar sites. Totally new systems such as the FST-2 were added to support the automated transfer of radar target data from the radar sites to the DC's. To support this added SAGE unique equipment at the radar sites new operations buildings were built. To go along with this growth in the foot print of the radar site was a growth in manpower. Some SAGE radar site manning increased to over 200. This meant more support facilities like barracks, medical, supply, and dining halls.

The new F-106 interceptors were entering service and were integrated into the SAGE system. To provide for digital communications connectivity between the SAGE DC, the radar site, and the F-106 some radar sites received new GKA-5 digital radios. They and the FST-2 were among the first digital systems fielded by the Air Force and relied upon solid state electronics. Maintenance now had to be able to repair older analog and newer digital systems, along with vacuum tube, printed circuit boards, and solid state (transistor) components.

Not all the changes necessary to transition to a SAGE radar site meant additions. Some equipment like the UPA-35 "controller scopes" used by operations became surplus since the SAGE system did the intercepts with controllers and the computer at the DC. These surplus items were sent to upgrade existing AC&W radar site capabilities and to counter new threats such as Castro in Cuba. Weapons Controllers at the radar sites were reassigned to SAGE operations at the DC's or to other manual radar operations in Europe and Asia. Even when older radar towers were replaced by new ones, they were left in place just to save money and provide more storage space. It was not unusual to see at least 2 old radar tower at SAGE radar site that were equipped with the new FD radar systems.

There were some unique problems associated with the new FD radars at SAGE radar sites. Some of these systems took years to work out the kinks in the systems. Others such as the FPS-24 search radar were massive. The radar tower was over 85 feet tall with 5 stories and a mezzanine. The rotating antenna system weighed over 85 tons with 100 horsepower motors for rotation. The antenna was 140 feet across and 55 feet tall. Problems with worn out antenna bearings became chronic. Replacing the old 8 foot diameter bearing with a new one required jacking up the 85 ton antenna about 2 feet. At Mt Hebo AFS, OR and Cottonwood, ID the weather (high winds and snow) made the addition of a radome to protect the antenna an absolute necessity. The radomes and their support structures were even bigger.

For SAGE, the communications systems were as important as the radar systems. A key component of a SAGE radar site was the Radar Data Processing System. This device allowed for the target information from search radars to be automatically sent to the SAGE computer at the DC. The SAGE system also included ground to air radios at the radar sites. These radios were connected to terrestrial communications circuits. The resulting communications path provided a means for the weapons controller at the DC to verbally direct the pilots of interceptor aircraft to enemy targets. As the SAGE system evolved, so did the capabilities of the communications capabilities of interceptor aircraft and the radar sites. The F-101 Voodoo and the F-106 Delta Dart interceptors included a computer that communicated with the SAGE DC computer. A digital radio was installed at the radar site to provide for the ground to air connectivity. The resulting communication between the SAGE computer, the radar site, and the interceptor aircraft was so automated, the pilot was only needed for take-offs, landings, and to pull the trigger if authorized.

SAGE speeded up the process by automating the flow of information needed to detect, intercept, and identify potential enemy aircraft. The destruction, if authorized would still fall onto the pilots.

There was much dialog about how SAGE got its name and what it meant. The most logical was that someone gave some sage advice and the brilliant people who heard the sage advice came up with the acronym SAGE first then created words to fit. There was no doubt the system was in fact semi-automatic because there was still human involvement associated with collecting height information and making tactical decisions. But the ground environment was somewhat confusing. There the most logical explanation was that the system was installed and worked on the ground and sent information to the air. No one really knows for anything certain about the naming of SAGE. But they knew the system worked.

SAGE was a computer, a big first generation computer. It took a half-acre of floor space; weighed 275 tons, used 55,000 vacuum tubes and each DC and CC had two of them. It took three million watts of electrical power to run it and it created an awful lot of heat. The SAGE computer automatically received and displayed to Weapons Controllers at a DC the azimuth (heading) information on aircraft detected by its' assigned radar sites. With the new system, the information received from the radar equipment at the radar sites was automatically transmitted over phone lines to the SAGE computer. The SAGE computer would identify it as either friend or foe from all the other known displays. All airborne aircraft were entered into the SAGE system. The SAGE computer also generated height information requests on aircraft. These were automatic requests sent to the radar sites where an operator would manually cause updated height information to be sent back. The SAGE computer would electronically display all the pertinent azimuth information collected onto a radar console CRT (Cathode ray tube- like a television set) at the DC. The SAGE operator of the CRT terminal used a light gun of sorts, to identify and display selected target information to include height data. The information collected and displayed about a target aircraft would be evaluated by officers and if deemed appropriate, interceptor aircraft would be launched. Once in the air, intercept information and directions would be sent from the SAGE computer to the autopilot of the ADC Interceptor aircraft. It would in essence take over the interceptor aircraft steering it to the target. The SAGE computer would not, however, be able to launch any weapons. Up to 150 CRT's could be supported by a single SAGE computer.

There were twenty-two SAGE DC's, each a large square concrete windowless building. These buildings were about 200 feet square and 50 feet high. Outside of each building were the air conditioning units, at least three of them all water evaporative coolers, and a power generator house.

Each center was fenced with one access point and manned by armed security policemen. Entry control badges were required to gain entry and had to be worn at all times while inside the blockhouse. Even with this security, there were various levels of security within the building. A scope operator would not have access to the computer room and the computer room people could not access the generators etc.

James Galloway had this comment about the internal security of the SAGE buildings.

As military personnel we were not allowed in certain areas of the SAGE building at Beale AFB. To be found in an area you were not authorized to be in was a security violation and the person might be disciplined. He or she might even lose a stripe or be fined over the incident even though they had a security clearance. That is, unless it was open house day. Then civilian personnel without any security clearances from the local community were invited in to see how their tax dollars were being spent. If a SAGE employee wanted to see what another section of SAGE did or what it looked like they had to disquise themselves as a civilian in civilian clothes. If in uniform and were caught in an area you were not authorized to be in you were still in trouble no matter that it was open house. I never could understand the rationale behind that.

The first SAGE Direction Center became operational at Syracuse, NY in 1959. This massive concrete structure was built within an urban setting and many people speculated on its purpose. Most people in the area thought the possibilities of nuclear war were more likely and the structure was to be used as a bomb shelter. They looked at it as evidence of America's willingness to fight if the Cold War became hot.

In addition to the twenty-two SAGE DCs there were three Combat Centers (CC) constructed. These were smaller than the SAGE DCs, but still windowless concrete buildings and all above ground. The separate SAGE DC and CC buildings were collocated at Hancock Field, Syracuse, NY, Truax Field,

Madison, WI, and McChord AFB, Tacoma, WA. These CC's had a different mission. They supervised the subordinate direction centers.

The North American Air Defense (NORAD) Command had been formed. This was an agreement between Canada and the US which still exists today. Canadian Air Force officers are stationed with the US at various military sites and have operational control over Canadian forces. Their actions are thus coordinated with the actions of our military for our mutual defense.

In the event a war broke out, NORAD regional CCs would have total control for the defense of Canada as well as our nation against air attacks. Later an additional capability known as Back-up Intercept Control (BUIC) was added to selected radar sites in the Continental United States which would act as SAGE centers in an emergency.

When AC & W sites were converted to the new SAGE methods of operations, AC & W sites became designated Radar Squadrons (RADRONS). As the conversions took place, the living conditions were improved. Facilities were transformed from WWII temporary building to modern quarters and recreation centers. Where weather conditions were extreme in areas like Alaska and Canada, and in some CONUS locations, covered walkways were installed between all site facilities to allow access to the various buildings without having to walk in or on top of snow banks.

Recreation facilities and equipment were made available. One great addition was the unique two-lane bowling alley that was installed at many sites. The alleys had manual pinsetters, which offered off duty airmen an opportunity to earn a few dollars by picking up the pins and placing them into the racks. The lanes were regulation and placed side-by-side. Space being critical these lanes usually occupied a narrow space next to a wall in some multipurpose building. Beginner bowlers often slip during delivery, and a bowling ball might go bouncing off a wall. It was not uncommon to see large dents in the sidewalls of these two lane bowling alleys.

These recreational improvements gave airmen many, new and different things to do when off-duty. Some sites had boats so they could fish. Skis, toboggans, ice skates and snowshoes were provided for winter exercise and adventure. Some sites had fishing camps where the personnel could go to get away from it all. Airmen were encouraged to organize drill teams and color guards and they marched in local parades and during open houses. Some had rifle and pistol shooting ranges and airmen were able to hone those skills. However, the main off duty activity was centered on the clubs. Not everyone used the clubs but they were a large source of distraction for most of the personnel and there were abuses.

There were instances of people becoming alcoholics and at that point the Radar Site Commander intervened and sent the man back to the United States where he could get treatment. In other cases, crimes were committed and those offenders were sent to an Air force Base for courts martial. After realizing this, Headquarters screened men selected for isolated and remote duty better to avoid many of these problems by rejecting those with alcohol abuse history. But the club was still the off-duty activity hub on most sites.

The club's were located on the site. Accordingly, individuals who developed drinking problems were easy to identify. The associated problems such as drinking and driving, and fighting were also easier to control on AF property. Usually, the airmen looked after each other and kept problems from getting out-of-hand. Rarely did an officer or NCO have to step in to maintain discipline. This made it easier for the site to be recognized as a good neighbor, and for site personnel to be welcomed when off-site. If there was any civilization near the site, airmen would patronize local restaurants and bars and were generally loved in spite of the rare minor scuffles with the residents. Then the site Commander would step in and make certain the airman was punished if it was deserved. Often this meant restriction to the site.

The other hub of activity on radar sites was the Dining Hall also known as the Mess Hall, Mess, or just Chow Hall. Four times a day, the Dining Hall was open and served food for the officers and airmen working shifts. Breakfast, lunch and dinner and what was called midnight chow, 2330 to 0030 for those people going on duty at midnight, and for those getting off duty from the swing shift. Some of the best Air Force cooks were assigned to the Radar Sites.

Dining Halls were and are operated on a budget. Most airmen eating in the mess were authorized to eat free known as being on government rations. Some airmen and NCO's received what was called separate rations. These were the personnel who lived off base. They had the option of bringing their own food or paying to eat in the mess hall.

All Officers received separate rations and would have to pay for each meal they ate in a mess hall..

Civilian employees did not receive compensation but were allowed to eat in the facilities.

When a person on separate rations or civilian employees ate in the dining halls they were required to pay a nominal amount for the meal. During the 60's, separate rations meant receiving $1.27 per day. Those on separate rations could either eat in the mess and pay for their meal, or if there were other facilities available, eat elsewhere. On isolated and remote sites there usually was no other place to eat so everyone used the dining facilities and if on separate rations they paid .27 cents for breakfast, .55 cents for lunch and .45 cents for dinner.

During this time, the daily budget for a Dining Hall was based on $1.27 times the number of personnel assigned that ate in the mess. So a 150-man radar site was allowed a budget of $190.50 per day. However if the radar site was a remote/isolated site outside the US, they were allowed a ration and a half per man per day so the same 150-man site had a daily budget of $254. A mess sergeant could supplement the daily food budget by using no cost emergency rations such as old C-Rations left over from WWII and the Korean conflict.

A single box of C-Rats weighed almost six pounds. A single case of C-rations were officially called ten in one rations, meaning it would feed ten men one meal or be enough to feed one man for ten days. These rations consisted of sealed cans containing bread, bacon, and a meat entrée, plus some condiments and a dessert usually sugar loaded candy bars. Most important were the cigarettes, instant coffee and the toilet paper. There were ten different entrée's available. The good ones, by that I mean palatable, were spaghetti with meat and tomato sauce; beans and sausage, and beef stew. Other entrée's like ham and lima beans were okay. There is no doubt that if hungry enough C-Rats would have been delicious.

Efficient Mess Sergeants were able to work wonders with C-rations. In many cases, the end result was as good as if the supplies had been obtained from the local supermarket…if there was one. He would open several cases of C-Rats and sort them putting all the beef stew into one serving pan, add spices and serve. Or he might separate the spaghetti into one serving tray and add some spices and make up some garlic bread and serve. A mess sergeant would not last very long if he just opened cans and dumped the contents into a serving tray. Again if you were hungry enough and perhaps blind it could be swallowed. However no one on a radar site could be that hungry and they certainly were not blind.

Efficient Mess Sergeants could also supplement his budget with fish caught by the airmen or with venison, moose or even bear hunted and shot by an airman. Another option was available to a designated Field Ration Mess at many radar sites. For reasons best understood by them, many airmen on government rations would not eat at the mess when not on duty (anywhere from 25 to 40 percent). However, the mess sergeant would still draw food from a government commissary for these people even if they did not eat. If they did not eat on site, the menu could be enhanced for those that did by using the "no shows" portion of the daily food budget. Most mess sergeants drew their basic monthly menu supplies from the nearby commissaries, but in the types and quantities selected by him. By using the "no shows" portion of his budget, the mess sergeant could add items over and above the basic menu such as pastries, steaks, and lobster tails. If every man assigned ate every meal the meals would have been hearty, but very, very routine. The key was being able to estimate the amount of "no shows" and was a "text book" example of Operations Research, or as some would say a WAG (wild ass guess). For example, by calculating the number of eggs the site should use and buying a percentage less, funds would be available for more expensive items from these commissaries. It only took a savvy mess sergeant to stay within his monthly budget. Good food was part of the high morale equation. At most radar sites, the mess sergeants were greatly appreciated by the officers and enlisted personnel.

The quality of food served in the Dining Halls was a great source of pride with the food service personnel. An annual award, The Hennesy Trophy, was made to the best Dining Halls.

Dennis Radke was stationed at Wheelus Air Base in Libya. He recalls that Cooks also had to be very crafty.

Local residents called Sadeeks, worked in the dining hall as mess attendants and clean up personnel. Routinely, frozen meat was brought out to thaw and because meat left unsecured was stolen, the meat was placed on a large table inside of a wire mesh area. Those precautions did not stop the theft. One day the thawing meat was steaks, a prized target for theft by the local populace and about half of the steaks came up missing. The next time steaks were brought out to thaw, the chief cook neatly placed a strip of bacon on each thawing steak. That ended the theft problem and the bacon was served for breakfast the next day. Often, to be a quality cook, you had to be crafty as well.

Throughout the radar sites, belts were pleasantly tight, and salutes given to the small, but exceptional productive and proficient food service crew.

So duty on an isolated or remote site had its drawbacks. The military man was away from his family for more than a year at a time. Generally speaking, there were no civilian females in the areas near the sites. The workdays were long and the weather miserable most of the time. In times of inclement weather, the troops on the hill might end up being on duty for days on end, eating and sleeping at the radar consoles until the weather broke.

Wind, heat, rain, snow, ice, all came with the assignments. The club, the Dining Hall, and the mailroom became the central point for most of the off duty people assigned. To keep morale and spirits high, cheap booze, good food and letters from home made these assignments tolerable. Most sites had Pool and Ping-Pong tables, theaters and hobby shops where leather goods, kiln dried coffee cups and ashtrays could be made. There were libraries where books could be borrowed and read. It was all designed to relieve stress. The stress of being isolated from their loved ones, watching radar scopes that they hoped would never have any blips being returned from enemy aircraft, knowing that if they did, they were a primary target. If targeted, they had no defensive rockets or anti-aircraft guns; their goose was or would be cooked. So they read mail, ate, worked, slept, played cards or sports, and drank.

CMSgt Edward D. Beard, USAF, Ret remembers the people of the 712th AC&W Squadron.

This is an example of the type of people we had assigned to our remote radar sites. I ran a small weather station out on St Lawrence Island (NE Cape) in the Bering Sea. As with most remote sites, we depended solely on aircraft to bring in supplies, mail and personnel. Our runway was a 3000' gravel strip about 1 1/2 miles from the main site. The Bering Sea has some of the most vicious weather on earth. A nice day is when the waves are only fifteen feet high. Our runway was lined with about 200 transportable lights. They were lined up on each side of the runway so the pilots could see where to land. My guys ran the lights and the homing beacon any time any air traffic was expected. These lights were always getting knocked out of commission by high winds and by dozers and graders clearing the runway. I arrived at the site in August 1957 to build and run a weather station at the runway. Every morning we would check the lights and report any outages to Civil Engineers. Often, the weather was so foul they couldn't get to them for a few days. In about mid-October, a civilian electrician named Gran reported in.

Gran was a retired US Navy Chief Electricians Mate. He had enlisted in the Navy as an underage teen-ager and served for 20 years. We knew his wife had died a few years before and he had two married daughters living near San Diego. Shortly after he got there, I noticed that any time the runway lights went out, he was there immediately to fix them. I can still see him out on that runway in the worst blizzards, hunched down working on those lights often with his bare hands. I would drive out there and try to get him to come in and get warm. But the Chief never quit until every one of those lights were working. Only then would he come in and get a cup of the rottenest coffee ever brewed by man. If you can picture winds of 60-80 miles per hour, snow so thick visibility was cut down to a few feet and wind chills well below zero that is what he worked in. He was always a pleasure to be around and regaled us with his many sea stories and chanties, many of which should not be told in mixed company. I can still hear him as he entered the club. "With a Ruddy F**king ho", and "Call the Captain, there's been some friggin' in the riggin'.

I don't know what happened to Chief Gran after I left, but I suppose he has long since gone to meet his maker. If so, I suppose God calls him first whenever any of his stars go dim.

During the Cold War, there were operational radar sites from the frozen outposts of the North American Continent to the sunny Gulf of Mexico, from the Atlantic to the Pacific oceans, and on many lonely isolated patches of mountaintops or mesas. The ADC people at AC&W sites, Texas Towers, Radar Squadrons, SAGE DCs and CCs, BUIC sites, and Fighter Interceptor Squadrons made our air defenses work 24-7, year after year. These dedicated people kept our enemies at bay and kept us safe.

Eventually the Union of Soviet Socialist Republics, the people whose Chairman said of the United States, "We will bury you" became the Russian Federation thus ending the cold war. But, during the cold war the resources of both nations were spent in almost limitless fashion developing weapons, keeping them operational, and training to use or defend against them. Then, because they learned the other side had them, they had to figure out how to protect themselves from those weapons. For forty years this seesaw battle took place. Lives were lost on both sides, of course not as many as would have been lost had the cold war been a hot war. Families were destroyed because of the assignments the soldiers and airmen, probably on both sides, had to endure and left their families home to suffer a different type of hardship. Some individuals became criminals by going Absent Without Leave, (AWOL) and some just committed crimes to get money. Some committed suicide and some went stir crazy. Some became alcoholics. But by and large, the Airmen of the USAF worked through their hardships and their loneliness. The improvements made to the isolated and remote AC&W's and RADRON's in the areas of buildings and of off-duty activities helped in the retention of Airmen and NCO's for second and even third tours and America remained safe.

THE PEACEKEEPERS

CHAPTER FIVE

The Duty

Each radar site had all of the same operational responsibilities as did large Air Force bases of the ADC. Every installation had to have a supply officer, a mess officer, a fire department officer, safety officer, etc. Each had to conduct inventories of sensitive equipment and accept sensitive communications. There were about fifty-three different areas which had to have an officer responsible. On sites there might only be eleven officers assigned so each officer had several additional duties and responsibilities. On an ADC fighter base there might be hundreds of non-flying officers, but still only the fifty-three jobs needing an officer assigned. So, there could a communications officer on duty at all times. There would be a food service officer whose only duty was to be responsible for the dining hall or halls. On bases there would be services such as laundries, theaters, stores and clubs that may only have an officer assigned as an additional duty but that may be the only additional duty the officer might have. A base might have thousands of people assigned whereas a site had about 150. On a site, most everyone knows everyone else. On a base there is less camaraderie, unless it is within the individual units. Medics would associate with other medics, security police with other security policemen, and so on. Entering a club for a drink, you would find the cooks and supply personnel at separate tables. They were still friendly towards each other but seldom would they associate.

As an example for on site operations, **Brian Coy** tells of his arrival at the 792 AC&W Squadron at Charleston, SC in June 1955:

It was a new site and getting ready to go operational. About 40 of us newly assigned airmen were met at the train station by the First Sergeant and a bus took us to the site. After being quartered we were led to the mess hall where a cook was behind the food line wearing a Hawaiian shirt and khaki pants. It was noted he looked a little old, almost middle aged.

After chow we went to our barracks and turned in. We were awakened by the First Sergeant the next morning and taken to the orderly room to process in. When the forty of us new airmen entered the orderly room we were shocked when we saw the cook from last night sitting behind the Squadron Commanders desk in uniform wearing command pilot wings and major insignia. He turned out to be one of the best. Major Edgar Armagost.

On the sites everybody was friends and everyone associated with each other. Security police, because there were usually only four to a site, associated with the one assigned medic, who associated with the five drivers, the six administrative, the two or three communications airmen, etc. Radar operations would have the bulk of the personnel with perhaps thirty with another thirty for maintenance. If they were talking shop they kept separate otherwise everyone mingled.

No matter what branch of military service, Officer Personnel had to remain professionally distanced from the enlisted personnel. On large installations, Officers always had their own place where they could let their hair down and relax without embarrassing their rank. Officer and enlisted normally did not fraternize socially with one exception. At Bases there are insufficient officers to field an all officer team, so teams will usually be built around organizations. There may be fifteen or more softball or bowling teams at a base each from a different unit. During these events everyone leaves their ranks on the benches. At small units, such as radar sites, there may only be one club. When it came to imbibing in alcoholic beverages, most officers were welcome into an all ranks club where they might socialize for a limited time and leave. When they left they were always sober, as they could never appear to be out of control. That would have been ungentlemanly and their conduct might be considered unbecoming to the officer corp. For that reason, even if there were only a small contingent of officers they built their own club or lounge.

At radar sites there was some fraternization between officers and the enlisted, certainly there was between the senior NCO's and the lower grades. This was out of necessity and was very beneficial at these small units. An emergency felt by one would be felt by all. Many times a family death would be met with a collection being made from the entire site to allow an assignee to get home for a funeral. If an Airman's wife had a baby while he was stationed at an isolated site, there would be a collection made to help. If a man got in trouble with the law, he generally was on his own unless he was being singled out. Then he would have the full support from his fellow GI's and from his site commander.

There is one story of an airman who had a brightly painted muscle car, but it got to the point where he could not afford gas so it remained parked on the site. That did not stop an over-zealous city policeman who had contact with the airman and his muscle car, and who saw a similar car speeding through town. The officer could not catch the violator but wrote the ticket and brought it to the site. The airman was served and appeared for his court date with the radar site Commander who testified that the car had not left the station. The Police Officer did testify that it could have been a similar car, and the case was dismissed.

Steve Weatherly, then a Second Lieutenant, related a story of a special event that broke up the mundane duty;

In 1965 while at Mt. Hebo, the NCO's requested a "Dining In" be conducted. It was scheduled for the same day that promotions would be announced. At these events everyone gets dressed up. It is a black tie event. Officers dress in what is called their mess dress, a snappy uniform with a white jacket, black tuxedo trousers, cummerbund, pleated white shirt and black bow tie. Military shoulder boards top off the jacket. The NCO's wear the regular air force dress uniform with a white shirt and a black bow tie.

I was a second lieutenant at the time and my rank insignia was a gold bar and it is what my shoulder boards would announce. As luck would have it, I was notified that I had been promoted to First Lieutenant. My gold bar insignia would now be silver. The problem was that I had no time to get the proper insignia. I could not appear at the dining in out of uniform. Being ingenious, I acquired some silver model airplane paint and gave the gold bars a coat. No one except the Commander knew of the "cover up" and in fact he used the same ploy when he was promoted from Major to Lieutenant Colonel.

James Hamilton tells this:

I was assigned to a radar site where the commander was a stickler for wearing a proper uniform. He would not allow fatigues to be worn off base and even specified footwear. Brogans had to be worn with fatigues and low quarter shoes with dress blues or khakis. Commanders on radar sites could be compared with the captain of a ship at sea. His word becomes law and his laws are obeyed. Although being a stickler at these small installations could be considered petty, there was an up side to this rule.

A local corporation used to collect toys at Christmas for the local kids. One day one of their pickup trucks loaded with boxes pulled up to the main gate with a tire that needed air. The truck and its two male occupants were allowed entry to the site. Both were wearing military uniforms under their civilian clothes. It turned out they were two penetration agents testing the security of the site. They were supposed to go around the various areas and plant small packages labeled "Bomb" to see if they could, then what the reaction would be when one was found and finally how the security plans worked if they were implemented. In this case they did not get far. He was spotted by an airman who yelled "Infiltrator" and other airmen grabbed the two men. They were wearing low quarter shoes with fatigues. Something no one assigned to the site would have done. The site passed the test with flying colors.

So by now, dear reader, you should have a fairly good picture of what life was like on these small remote and often isolated radar sites was like. The work was performed by dedicated personnel. Operations searched the skies for any sign of attack by our cold war enemy knowing that if it occurred they probably would be killed. They were supported by other Air Force personnel who because they were at the same location would also perish in an attack. Still they did their jobs. They worked hard and they played hard. They looked for things to do to occupy their minds. When minds were not actively working or actively playing they turned to other distractions. Fishing, hunting and other outdoor sports were common events among the men. However, some were just not athletic enough to enjoy the outdoor activities. Many men received higher degrees through correspondence courses. Card games and drinking were distractions during the long winter months and inclement weather, but not the only one. A major distraction for many was playing tricks and jokes or each other and in some cases getting even.

PART II

SOME OF THE MEN AND SOME OF THEIR STORIES

The Peacekeepers

THE PEACEKEEPERS

CHAPTER SIX

This was basic training

Basic Training for the armed forces is designed to take civilian boys and girls and turn them into military men and women. At least that was the intent at the time when most of the personnel who were assigned to radar sites went through it. Basic was from four to sixteen weeks in length depending upon which branch of service was conducting it and when it was. This was the time when the recruit was issued his uniforms and taught how the military operates. What the expectations were and how the recruit would be expected to satisfy those expectations. The recruits learn the proper wearing of their uniforms, how and when to salute and how to move as a unit from one place to another. In the military it is called marching. They are taught how to keep their living quarters and they are taught when they are to eat and sleep. Their days of being an individual are put behind them and they are taught how to act as a member of a team.

Ron Dotson on staying out of trouble:

I am reluctant to write this for fear that someone involved may not be over something that happened 54 years ago.

In August and September of 1956, I was in Basic Training at Lackland AFB, Texas. We were in the new barracks at that time and were housed in four man rooms. One morning we were getting ready to fall out to march to chow. One of my roommates and I found that while we were in the latrine some low life had stolen our flashlights and arm-bands. We quickly ran over to the other bay and stole someone else's.

By the time we reached the assembly area in the pitch-blackness of early morning the Flights were falling in. Like faithful dogs we could only find our flight by hearing our TI's voice which we ran toward at full speed. Coincidentally a flight between us and our TI's voice had just been given the command "Right face - Forward march". Unfortunately they had not turned on their flashlights. We hit the four squad leaders and right guide with a beautiful open field block and took them out. The whole flight of 40 men folded like an accordion. Simultaneously their TI began cussing at the top of his voice. My friend and I jumped up and ran around the opposite side of the jumble of bodies and away from the screaming TI. We quickly found our flight and gained our composure as we marched to the chow hall wearing armbands and with flashlights.

We were marching back to our barracks about 30 minutes later in the early dawn light. As we were passing an athletic field, there was the flight we had destroyed.

Forty young men were crawling around on their hands and knees mooing at the top of their voice. The TI was standing in the bleachers yelling "You people can't march. You can't even walk. All you can do is mill around like cattle and graze. You people ain't gettin' no chow this morning."

I still have pangs of guilt but then I think about how ridiculous the whole thing was and we got our chow that morning. If anyone reading this was in that flight, please accept my sincere apology.

Bud Irwin, Lt Col (Ret) tells of his training days at Lackland AFB as an enlisted recruit:

I enlisted 4 August 1950, soon enough after the Korean "conflict" started in June, to still have a two story WW II barracks to live in. However, it wasn't long before the flood of new people exceeded the capacity of Lackland's buildings and most basic trainees were required to live in tents. According to one of the letters I wrote home, there was a day while I was there when 2,000 new troops arrived. The country was woefully unprepared for Korea and the USAF was short of everything, including people. The 1947 conversion from being a part of the Army to being an independent service appeared to be still in progress. The blue dress uniform had been adopted and the summer uniform was officially khakis, as before, but with a blue tie, black shoes, and dull silver insignia. There evidently weren't enough to go around because we were issued standard Army khaki ties and brass insignia (that had to be shined) and wore the Army Air Corps patch on our sleeves. We generally wore this uniform only for inspections, our normal dress being one-piece fatigues with a belt around the middle, a helmet liner, canteen, and brogans. Not all brogans were the same, they issued whatever they had left over from WW II, and mine were a type that I found to be the most comfortable footwear I've ever worn. They were made with the smooth side of the leather to the inside and the rough side out and they had a soft toe. The only bad part was that they had to be shined and getting that rough leather to the point where it would shine took a lot of work. We washed our own underclothes and fatigues nearly every night. This was done by wearing our underwear into the shower, putting our socks on our hands and then scrubbing down with strong G.I. lye soap. Our fatigues were laid out on the shower floor, scrubbed with a stiff bristle brush and soap, and then hung over a showerhead to be rinsed and wrung out. I had learned to march in high school ROTC and the Flight Chief made me right guide for my flight. This placed me third in command, in charge of the four trainees designated as squad leaders and their 65 men. I was required to march at the right front of the flight and set the cadence. I also had to prepare guard rosters, march the men to various formations and classes, generally assist the Flight Chief and his assistant in managing the flight, and fill in for them when they weren't present. The good part was that I didn't have to pull guard duty or K.P. About the end of August, the Assistant Flight Chief was sent to Korea and, since no replacement was available, I was made an "acting" Corporal (although still a Private) and given his job. Like I said, the USAF was really short of people. This gave me the privilege of no longer waiting in line to eat but really increased my workload since I had to do that job and also attend all of the training classes and other events required for a basic trainee. Near the very end of our six weeks of basic training (reduced from eight about when I got there) we were moved into tents to make room for an incoming flight, the Flight Chief either got a new flight or shipped out for Korea, and I took over his position until the flight departed for their next assignments. I guess it isn't stretching things too much to say that I took my own flight through basic training. Incidentally, I skipped a year in school and was only 16 when I enlisted.

Pete Holland who made the USAF a career and is a member of the Air Force Radar Site Veterans (AFRSV) provided the following:

Just before graduating from high school, my Dad and I went to visit the Air Force recruiter. I was still 17-years old and needed permission to enlist. Dad was a career US Army Military Police First Sergeant. I knew that I didn't want to be a soldier. I enlisted in the USAF and became part of a "Buddy Flight" made up guys from the south central part of Pennsylvania, who would go through basic training together in the same group or flight. We were all inducted and took the oath of enlistment at the same time at a Memorial Day celebration in the town square of Gettysburg, PA, with parades, Army Fife and Drum Corp, etc. We flew out of Baltimore, MD to San Antonio. The local newspapers followed our progress through basic training and wound it up with a big promotion. Even the recruiters came down to Lackland for photos and glad tidings after graduation. While waiting for the flight, one of the guys hooked up with me. My Dad told me that this guy would not make it through basic training and he didn't. He lasted three days before being transferred to a Casual Flight for out-processing. We lost about five guys the first week or so.

I never witnessed any physical abuse during Basic, but there was plenty of psychological abuse to go around. It seemed like the TIs were in your face constantly during the first couple of weeks. Probably to weed out those not suited to the stress of the military. One thing that really ticked me off was that the TIs would set you up. If they couldn't find a laundry tag on a uniform during room inspections they would put one on themselves and give you holy hell for not removing it. After the first couple of weeks things were still tough, but tolerable. After the graduation parade, I thought I had it made, as brand new Airman third class (A3C). When we got back to the barracks, I got a coke out of the machine and was just about to take a long pull off the bottle, when I heard,

"You're not out of here yet, get rid of that bottle!" I'm sure that the TIs were having a good chuckle over that one. That night we ended up being sent to tech schools all over the place. I boarded a bus abound for Keesler Air Force Base Mississippi. And so my adventure began.

John Kimmes related this about basic training:

I was young and only had a little "peach fuzz" for a beard but I was told to shave every day whether I needed it or not. I was several days into basic training when during an inspection the TI accused me of not shaving. He wouldn't believe me when I said I did shave. So, he told me I had 30 seconds to shave and get back in formation. I ran to my room, got my razor and shaving cream, and headed for the latrine. It was then I discovered I'd been shaving with the fake cardboard blade still in the razor. Talk about getting red faced!! Anyway, I put in a real blade and proceeded to do a fine job of nicking up my face. It took more than 30 seconds but when I returned to formation I had some additional red on my face. The TI had me put my hands together and he poured out some nasty stuff and told me to put it on my face. Lord, did that burn!! It was almost like that scene from the movie "Home Alone". That happened back in 1965 and I still tell story today for a good laugh.

Allen Miller, who retired in 1978, relates his basic training experience:

I enlisted in March of 1949 and had to take thirteen weeks of basic. We had a flight and squadron number, of course, but I can't remember either. I do remember the flight chief. He was a Texan; he had three stripes; he wasn't with us the first week or two of training because he was recovering from a broken leg that he received when he parachuted out of an aircraft. Why did he jump out of a plane? Because he wanted try it at least once in his life, that's why. His name was Sergeant Comer and we all envied this guy because he would often leave us under the care of the assistant flight chief while he went off with one pretty young thing or another; they all seemed to drive new convertibles and always had the top down when they picked him up.

But one particular weekend he did not abandon us. No, he stayed around the entire weekend and supervised the most thorough GI Party ever. A GI Party in those days was where we all got together and, on our knees, scrubbed the unfinished wooden floor of our barracks. Sergeant Comers goal was to have a white floor.

We started this particular party right after dinner on Friday. Sgt. Comer had the first floor, the assistant the second. We moved all of the double bunks to one side of the barracks. Footlockers and shoes were placed on the lower bunks. Those of us who did the scrubbing lined up at one end of the barracks almost shoulder to shoulder on our knees. We were wearing our shorts and each was armed with a two-gallon galvanized bucket of water, a new bar of brown Fels-Naptha soap, and a stiff bristle scrub brush. We were instructed to scrub the area in front of us using one-foot strokes -- one foot forward and one foot backward, one foot forward and one foot backward. We were to stay in line scrubbing that one area until Sgt. Comer blew his whistle and then move backward one foot.

Comer was sitting at the end of the top bunk in line with the row of scrubbers with his legs hung over the end. As we scrubbed he would call out to particular scrubbers that they should bear down harder or that they not forget the area to their left or that they didn't have enough soap on their brush. When he was satisfied he blew his whistle and we all moved back exactly one foot. He dressed our rank instructing individual scrubbers to move forward or backward. When we were properly lined up we commenced scrubbing -- one foot forward and one foot backward, one foot forward and one foot backward. Comer himself would move to a new bunk when we got too far away from him.

As soon as there was room for them to work other airmen with mops swung into action mopping up the soapy water we left in our wake. They were equipped with heavy GI mops and mop buckets fitted with wringers. There was a seemingly endless stream of men replacing the mop buckets filled with dirty soapy water with buckets filled with clean fresh water. They'd put down the bucket of fresh water, transfer the squeezer to it, and carry the soapy water back to the latrine and empty the contents, rinse the empty bucket and refill it with fresh water.

 When we finished the first side of the barracks we transferred the bunks to the other side and started working on the other half of the floor -- one foot forward and one foot backward, one foot forward and one foot backward. We couldn't wait to finish that side. We had no idea of what was in store for us. We all had assumed that when we finished we were done for the day. We were wrong. When we finished we started scrubbing again. This time it was the walls, windows, posts, even the top of the heating vents. We kept this up till perhaps 2130 hours, a half-hour before lights out. In that half-hour we relocated bunks and footlockers and had time enough for a quick shower.

On Saturday we re-scrubbed everything between breakfast and lunch, between lunch and dinner, and between dinner to almost lights out. On Sunday it was the same thing again except that churchgoers were given time to attend Sunday service. Needless, to say, everyone got religion those Sunday mornings and took a break at church.

We also had a man on duty between lights out and reveille acting as fire watch. I recall it being a one-hour tour. But we didn't just "watch." We spent the hour scrubbing the stairway to the second floor. This went on for the entire balance of basic training under Sgt Comer.

At sometime on the first Monday following our scrubbing marathon our barracks was inspected by some high-ranking officers. (We were off somewhere on a training exercise led by the assistant.) The story was that they were so impressed by our beautiful white floors that they invited their superiors to take a look. It was rumored that the Commanding General of Lackland AFB looked at our floors.

I do know that Sergeant Comer was the proudest flight chief you ever saw. He delighted in telling us how the inspecting officers had oohed and aahed over our floors and how one officer had said, "And if they ever let those stairs dry, I'll bet they'd look as white."

Bob Irving tells this of his basic training:

On the first day we were changed from Rain-bow (because of the various colored civilian clothing the recruits wore) to "A-1" looking Air Force Troops in our new uniforms. That night were had to Shine Shoes until revelry, but we were told don't get caught being up past lights out. Our Training Instructor (TI) required us to stay in our Barracks.

During the first week we had to learn the ropes, all the rules and be able to repeat them on the drop of a dime. We were taught how to set up footlocker and how to align our hanging items. We learned how to make three types of beds, stripped, with the mattress rolled up, blankets folded and pillow squared. A white collar for inspection with the blanket, and sheet folded to exactly four inches wide and a specific number of inches from the top of the bed, and finally, the normal everyday make-up. Our beds were made with the extra blanket used as a dust cover for the pillow. The blankets had to be tucked under the mattress with "hospital corners." This meant every corner had to have exactly a forty-five degree fold with the blankets tucked tight enough so if a coin was tossed onto the blanket it would bounce. During this first week we also were taught how to correctly wear our various uniforms and most importantly exactly where to place our collar brass and nametags.

During week two we had classroom training, physical training and marching, marching and more marching. We also had a fire drill in which we were told to exit the barrack with much exuberance. When the bell sounded we did as we had been told so much so that we accidentally ran over the training instructor supervisor in our haste. Later we found out our TI didn't like him. Once, I was being chewed out it was like what a movie spoof might be. I was 6'4" tall and our TI was much shorter. He grabbed the front of my shirt and pulled me down so we could be face to face. I could not help but start to laugh. Our TI just walked away shaking his head. I assume he didn't want us to see him laughing as well. From that point on basic training became more fun for me. It wasn't for everyone. During an open foot locker inspection one guy failed so badly, the TI emptied a can of shaving cream into it then shut the lid. He then made the poor guy run up and down the aisle yelling "oink, oink" until he cried. Needless to say he was one of many that didn't make it.

During week three we had more classroom training mostly on the history of the USAF. Our marching and physical training was conducted between the barracks in the shade because of a red flag being raised due to the heat after all it was the middle of August and we were in Texas. We still had to march to and from the obstacle course, and to the gas chamber for more fun and experience with tear gas. By now others were having fun as well, in fact so much fun, during week four we went through the obstacle course again. We were so tough we marched to it while others took a bus. All of our marching and drill instruction paid off when we won the marching competition. As I was the tallest in our flight, I was the right guide. That position is on the front right side of the formation and everyone else uses that as the position to form on the left and to the rear so that the formation looks square and military. Being the right guide, I was told if I screwed up, everyone screws up. Talk about pressure. The rest of our training was uneventful.

Upon graduation we got assignments to an advance training school. My orders sent me to Keesler Technical Training Center. I was going to attend the Radar Operators School and would be affectionately known as a Scope Dope.

Gary Hagan had some memorable experiences:

I enlisted 1 Dec. 1959. One memorable experience I recall was a night march. We had a lot of fun crossing a pond on wet logs. Many of us got a little wet. We also had the "smoke house" experience. With gas masks off we walked in the front door and waited until the tear gas popped and until our eyes felt like they were burning out of our heads. Then, and only then, the TI said, "Mask On." We fumbled around and got our masks on, cleared it, and uttered some phrase to the TI. That's when we were allowed to clear the house and rip the mask off and try to get our eyes to work again. Another time around the "confidence" course, I came around a little hill and stepped into a rut. My ankle turned out and I heard it pop. I fell to the ground and then realized that if I didn't keep going I would be "phased back." Not wanting to start basic all over again, I pressed on. That night after the rest of the guys were asleep I napped in the shower, sitting on the floor with cold water running on my ankle. My ankle had reduced its swelling to the point that I could once again see an ankle bone. I was the first one finished shaving next morning, when that scratchy record player started up sounding reveille. I dressed and really tightened up the laces on my boot which made a walking splint and allowed me to make it through the day.

I learned a lot about myself and human nature during those six weeks.

John Ferlaak tells of his time:

When I first enlisted, back in Oct 60, I was living in San Clemente, CA. The recruiter wanted to know if I wanted to be part of a "special Flight" called the Fullerton Flight leaving 29 Oct 60. What this turned out to be was a flight of about 50 recruits, from the same area (Orange County) all being sworn in at the same time and would stay together through Basic. They (the recruiters) had a big event planned, with special guests, lunch, pretty girls (models) and the swearing in ceremony. Everything went fine until we arrived at the airport to board our chartered "Super Connie". Fog rolled in and delayed the aircraft until about midnight. So we had to find ways to keep busy until then. Don't ask why, but I had brought along a "Pewee" football, which we used to occupy some time just passing it around, until the plane left at mid-night.

We arrived at San Antonio about 0500 and bussed to Lackland Air Force Base, (AFB) Texas arriving at sunrise. When we got off the bus, we got our complementary "yelling, cussing, and name calling", followed by the ever so popular breakfast of runny eggs, cold greasy bacon and SOS. We were then told we would "march" to our barracks, where everything we owned would be confiscated or sent home. Somewhere about half way across Lackland, that Peewee football landed in a drainage ditch for fear of being confiscated or watching its owner being verbally reduce to something lower than whale poop or worse.

Thinking that we would all get some rest, after the long flight and being up all night, we soon found out that we would be "processing" all day, since our Flight was late getting in. It felt so good to finally hit the sack at 2100 only to be awakened at "zero dark 30" for a fire drill. Finally we were able to get back to bed until 0430 when we awoke to finish our processing. I know I'm not the only one who wondered what we had got ourselves into.

It was nice that all of us were from the same area and had common interests, but it did prove to be a major problem getting the flight to come together as a well disciplined group. The TI's had their "hands full" with this group and wondered if we would ever pull together. We always got a big cheer from everyone, when someone would ask us if anyone was from California. The worst TI was an assistant "one striper" called Vorhees. Talk about somebody on a power trip! To this day, when I see that name, I wonder if it could be him or his relative. What a piece of work!

Jim Saunders related how they made the meager pay of a basic airman stretch:

When I was in Basic at Lackland I was in one of the Disneyland barracks. We had a 'patio' nearby that had drink machines. One of the machines would give you a soft drink without putting any money in it. Just push a button. Well, the word got out. There were guys lined up with their laundry bags filling them with drink bottles until all the drinks were gone. Our TI found out about it the next day and warned us not to do it again. That was all that was said.

James Hamilton tells of some of his time:

I joined the Air Force in June 1949 with several buddies from High School. After I had passed the physical, I came down with a severe case of tonsillitis and running a fever when I boarded the train to get to Lackland AFB. My buddies felt I was a handicap to their having fun, and Charles E. Jones, a black recruit volunteered to take care of me even though we did not know each other.

We finally arrived at LAFB and although still sick, I and the others went directly to the dining hall where we were served our first military meal, cold, very salty, navy beans and bread. Next, we were directed to the supply room and received our bedding, and then to a two story wood frame building that would be our home for the next thirteen weeks. A TI showed us how to make a bed the military way. He showed us one time then tore up the bed. For the next week everyone thought that getting their beds tore up after they made them was part of the ritual because it happened every day even if it was made properly.

We were marched everywhere when we really did not know how to march, but we were learning. Sometime after the first week we were issued our Air Force uniforms, underwear, socks and shoes and boots and taught how to dress. We had become Airmen recruits. A regular TI and a Drill Instructor (DI) were assigned to us. Their education concentrated on there being only one way to do things, the Air Force Way. TI's made life miserable for the recruits and many had fun at our expense. If an officer came by and was not saluted properly, the recruit would be ordered to stand in front of an outdoor mirror saluting his own image. This would go on for a long time or until another officer came by and was properly saluted. Then the recruit might be allowed to go his way. If a recruit did a left face when the order was for a right face, the stupid recruit who did not know his left from his right would be given a rock, usually about three pounds or so, which he would have to carry in his left hand for the day or until told by his TI that he could put it down. He could only have this rock in his left hand so he could still salute. It would not be unusual to see a recruit holding a rock and saluting a mirror.

Recruit barracks were inspected very rigorously by the TI and occasionally by another NCO. To prepare for these inspections which were held every Saturday morning, on Friday night there would be what was called a "GI Party."

The floors were scrubbed, waxed and polished. The windows were made squeaky- clean. Each individual recruit made certain his wall-locker was appropriately laid out with dress uniforms on the left and work uniforms on the right. Every hanger positioned the right way, the Air Force way, and evenly spaced one inch apart. Footlockers were identical with a place for everything and everything in its place. Every bed was identical and tight. Any coin dropped on the bed had to bounce. The barracks was washed and dusted upstairs and down. Every recruit properly dressed had to be standing at attention by his bunk, under which were neatly aligned his spit shined boots and shoes. You could stretch a string from one end to the other and everything would be aligned. The inspector would enter and the command "BARRACKS ATTENTION" would be yelled by the first person seeing the inspector. The inspector would look up and down. If it looked like he could find nothing wrong, he would remove from his pocket a white glove and put it on. He would then begin wiping the glove over various surfaces until he found dust or dirt that someone had missed, and there would always be some somewhere. The recruits would be on restriction for living in a pigsty.

For the barracks that I was in, it had been seven weeks of restriction because of the filthy living conditions we lived in, according to the inspector of course. You could have eaten off most any part of the barracks. After all, it was Texas, and the wind blew, and the sand or dust would be blown in overnight so we had suffered for seven weeks. There had been no relaxation, no movies, no beer garden, no service club privileges, and no Base Exchange, except to get a haircut or to pick up laundry, and immediately return to the barracks. It was the Friday of our seventh week and time for our GI party. We knew every nook and cranny in the barracks where a white glove had come away "filthy." We made everyplace sparkle paying special attention to those nooks and crannies. We cleaned upstairs and down like they had before and they had everything aligned perfectly. I was detailed by our TI to get the Charge of Quarters from the orderly room and to be prepared for another white glove inspection. I did as I was told and on the way back to the barracks, the Charge of Quarters asked if we were ready for inspection. I told him he would not find even a single piece of dust because of all the work we had done. It was the wrong thing to tell the inspector, he probably took it as a dare. When the inspector entered the barrack which was at attention, he looked and he smiled at how perfect everything looked. He removed his white glove and stood on one of the bed frames. Then he slid his gloved hand up the wall near one of the framing timbers to where the second floor was attached, a place they had never seen inspected before. This was where dirty water from the second floor would run down to the first floor. He pulled back a glove that was wet and black. He walked around the first floor holding his gloved hand up for all to see and restricted us for another week.

I never told any of my barracks mates what I told the inspector. I was afraid they would have killed me.

David Norrick reported that one of the things he did during Basic Military Training in 1966 was to be assigned to guard duty:

Now, one would think guard duty meant the guard would have a weapon. BMT guards were armed with a flashlight and their most important role was to yell fire if they saw one. In those days all of the barracks were of the two story wood frame buildings each housing about 60 men. Fire department test revealed that once lit, that type of barracks would burn to a pile of ashes in about seven minutes because they were all built during WW II and the wood was extremely dry. So as I diligently patrolled my area I saw a pair of headlights pull into an area between the barracks and shut off its lights. My instructions were to challenge anyone on or near my post, so I approached with his trusty flashlight on hoping the occupants of the car would see or hear me and leave. They did not. I knocked on the window and following my instructions, said "Halt who goes there?" I was surprised to see the face of my Airman First Class assistant drill instructor and a young lady. The drill instructor said, "Troop do you want that flashlight up your arse?" "No Sir" was the reply and the Instructor then said, "Okay, carry on." I again did as I was told and continued on my guard duty laughing to myself every time I passed the car. The next day I was called into the Drill instructors dayroom and offered the position of flight leader, which meant no additional duties. As enticing as that was, I respectively declined relying on the old military adage, never volunteer.

Guy Bierwirth Sr. On the shooting range:

During training on the rifle range we were told to button the top button of our fatigue shirts to prevent a hot shell after being fired and ejected from the weapon of the man lying next to you from going down the back of your neck. Experience had taught the instructors that if a hot shell did go down the collar of a man he might do many strange things. We had a "simple" farm boy from Utah who got a hot shell down his neck. He turned around and in doing so his loaded weapon was directed at all of the TI's. He was jumping up and down trying to get the shell off his back and was met by a TI who grabbed his weapon, put the safety on and threw him off the range.

He was set back for one week to another flight. However, the instructor failed to do the proper paperwork regarding the transfer. The Squadron carried him as AWOL for several days until they finally figured out what happened. So, to cover their behinds, they blamed it on him and set him back another 2 weeks. We lost track on him and never found out if he made it out of basic or not.

Ed Beard, Running the pay gauntlet:

We were taught how to report for pay. We were taught to verify the amount we were paid against the master payroll. We were taught to always count the money before leaving the pay table in front of the pay officer. We were taught once we took our pay and turned away from the table we could not claim there was an error. We were not taught or even told of the gauntlet after we left the pay table.

I reported for my first month's pay of $58 and noticed there were a lot of one-dollar bills but they spend just like real money. I verified that the $58 was the balance of my first months pay after they deducted the flying twenty we had to take to get hangers, shoe polish, marker kits, towels and a who bunch of stuff we were told we needed for our footlocker and that we could never use. I counted the pay and signed the pay roster, saluted and turned away from the table like we had been taught. Then I ran into the gauntlet.

The gauntlet was a row of tables. Seated behind every table was a high ranking NCO from our basic training squadron. As required, I stopped at each table and reported to the senior NCO at which time I was informed he was collecting for the Red Cross or the Old Soldier's

101

home or the Air Force Aid Society or any number of other charities. By the time I had run the gauntlet I was eight dollars poorer than I was when I collected my pay. However I did receive a benefit. A lady was seated behind the Red Cross table. Next to her was our first sergeant who told each of us that with a proper donation we would receive a card that would entitle us not to be placed on the KP roster. I donated three dollars and did not receive a card. Another recruit, who apparently donated more and received a card that the lady failed to put his name on. Sure as hell several of our names appeared on the KP roster but we knew our buddy had the card. One by one we went into the orderly room, reported to the first sergeant, showed him the card and the first sergeant struck our names. As we left we gave the card to another buddy who was in line and the process was repeated. By the time the third man showed up saying the same thing, that the lady did not put his name on the card, the first sergeant caught on. We should have got at least a good mark for ingenuity but instead we were put back on the roster and had to pull some extra duty besides.

Larry Burfield on his introduction to the USAF:

My first day at Lackland AFB on 16 January 1966 had to be one of the worst. To start off the day we arrived at the San Antonio airport around 12am after a fun flight from Port Columbus (Ohio) we were greeted by the typical foul mouthed drill Sgt in training. This wasn't so bad for me because I was used to marching around and getting yelled at in High School. I had to try and not smile at this show of buffoonery, after all, he was the one who had to get up and meet us idiots at "zero dark thirty" in military time. The bus ride to Lackland proper was nothing to remember except on arrival we were ordered to bail out of the bus in the shortest time possible and we were pushed along by three more DI's in training (I think the Real DI's were still at the NCO Club drinking).

In their effort to push us all out of the bus several of us were not able to grab our small travel bags from home and they were left on the bus. Our attempt to explain this to the Sgt was met with the response you would expect from a guy whose job is to yell, scream and insult your family tree. We never saw the bags again and I expect the bus driver has a pile of these bags somewhere in south Texas to this very day. That was the good part of the day.

Later in the morning, around 5am, we were told to turn in and get some rest. This was simply an effort to get us off guard. Promptly at 0530 the fire alarm went off. The fire alarm was the 'real' Drill Sgt (the NCO club must have run out of booze or closed) banging on a trash can with a "billy-club" and his assistant Drill Sgt's turning on the lights and shoving people out of the bunks while yelling "FIRE! FIRE!" at the top of their well-developed lungs. We were ordered to run outside and line up in ranks as fast as humanly possible. We had just done this a few dozen times and were pretty good. The assistants were supposed to hold open the doors to let the mob exit the WW II barracks. I was upstairs and so had to run down the stairs and out the door. Trouble was the Drill Instructor let go of the door just as I was careening down the stairs at full gallop. I crashed through the door with my head leading the charge. Glass and wood splintered, the Drill Sgt was knocked on his ass and I kept on going, as ordered, standing at stiff attention in the ranks with the other airmen basics. Trouble was I was bleeding like a fire hose and when the DI came up to chew me out he took one look and was kinda lost for words-for a split second. He went into a full blown hissy fit about me bleeding all over "HIS" newly issued Air Force field jacket and why the hell was I wearing it anyway?

I was a little confused about whose field jacket I was wearing when I realized what he was saying was that he as a taxpayer owned the field jacket and I was only taking custody of it on a temporary basis. I started to explain that since we had done this fire drill business several times, I had just grabbed it on the way out. His face was turning redder than my blood and his eyes were starting to bug out in rage. Trouble was I had grabbed the wrong field jacket and was bleeding all over some other poor guys. They wrapped a towel around my head and one of the assistants drove me to the emergency room to get the glass removed and stitches put in, then it was back to the barracks and on with basic training. I didn't even miss a day and the DI never said anything about me knocking him on his can. His temper never improved. Still have the scar on my head; don't know what the other guy did about his bloody field jacket. The rest of basic was pretty normal.

Having been working on a horse farm in Ohio I was ready for the physical part, in fact it was pretty darn easy compared with farm labor. I kept winning push-up competitions and that sort of thing. It always came down to me and this really big guy, but I always lasted just a little longer or did one more chin up. He could have beaten me in weight lifting I'm sure. The "reward" for winning was always the same, a smoke break. Trouble was neither I nor the really big guy smoked. We always had a good time with the contests anyway.

THE PEACEKEEPERS

CHAPTER SEVEN

Technical Training

Training for most of the technical jobs on a radar site was conducted at KTTC. The training period was from four months to a full year depending upon what particular job an airman would be assigned. For the most part the first week of being assigned to KTTC the new airman would be assigned KP for a week. After that his training would commence.

Trainees performed some of the tasks normally performed by TI's and DI's, tasks, such as marching the new airmen to and from the mess hall, to and from their training schools and supervising the cleaning of the barracks for inspection. These trainees would be identified by various colored shoulder cords called a fort-a-guerre, and referred to as a "rope" which they wore on their right shoulder. Usually, a red rope was responsible for the barrack floor where they lived. There were two such red ropes, one for upstairs and one for downstairs. An airman from that floor with a problem, or who needed supplies, would go to the red rope for a resolution. Another airman was given a yellow rope and he was in charge of the entire barracks. A white rope signified the man in charge of the entire squadron or group of barracks. Ropes did have some authority but only over the airmen assigned to them.

Technical training was conducted at many different locations on KTTC and the training was performed by either skilled military personnel or by civilian employees, both expertly qualified to instruct.

105

It wasn't all work and fun and games. Quite often there was boredom and with little money there were not too many things to do so they invented.

Don McDermott passed along a story of how some entertained themselves and caused some concern among the Air Police:

At that time the barracks were heated by gas furnaces. Some enterprising airmen would fill a condom with gas which of course is lighter than air and which would cause the condom to act like helium filled balloons. They would tie it off with thread and soak the thread in lighter fluid. Letting the balloon rise to the end of the thread, they would light the thread and let go. The balloon would rise, the fire would climb the thread and when the fire hit the bottom of the balloon, it would explode. Naturally there would be reports to law enforcement about the explosions. The Air Police would react thinking there was some sort of attack. Zeroing in on the source they would only find airmen diligently studying their training manuals quietly laughing among themselves.

Patrick Smith tells of the time he was assigned to Kitchen Police (KP)"

Typically there is a lot of grumbling about KP and rightfully so. The days are long, and the work is hard and mostly wet and dirty. However there can be some side benefits. In 1961 I was assigned KP over Thanksgiving. I and three others were assigned a heaping thirty-gallon container of cooked shrimp to shell. One of the cooks, because it was Thanksgiving, brought us a container of shrimp sauce saying it might make the task a little more enjoyable. It certainly did, however the cook was not too happy when we gave him about three-quarters of the container back. However, what the Air Force lost in shrimp they probably saved in turkey because none of us KP's ate much of that during dinner.

A by-product of all this cooking and feeding is garbage. Three of us were chosen for "garbage detail" as our personal contribution to the academic endeavor. Our role would find us entrenched in a malodorous, central courtyard at a place called the "garbage rack" which adjoined four enormous chow halls. Within each chow hall resided its own dishwashing facility, or 'clipper", where prior to washing the countless pots, pans and dishes, they were relieved of their leftovers. These leftovers morphed into garbage yielding immense quantities that interminably funneled into our courtyard.

Imparting to us in the most unambiguous of terms, the Mess Sergeant communicated precise instructions for our part of the KP adventure. Our job called for ceaseless sorties to the clipper areas where we would capture and return brimming cans of garbage to our courtyard. There we would empty the contents into a large container, then sanitize the empty cans on a steaming apparatus, and once restored to a glistening state, return them to the clipper rooms. Then, repeat as often as necessary, which meant it was ceaseless.

As chance would have it, while milling about in our courtyard purposely avoiding our task, one of the KP's from one of the clipper areas walked out their door and yelled, what do we do with our garbage? Seizing upon the apparent fact that those in the dishwashing areas had not been told of our part in this effort, the Mess Sergeants orders faded into obscurity and were abandoned. With nary a beat of remorse, we instructed the airman that he was to lug the cans to our area. Kismet had smiled upon us! Quickly monopolizing upon the unexpected fortune, we soon had all the clipper personnel schlepping their own heaping cans of muck to our courtyard, emptying them, steaming them and carrying them back. Our contribution…well, it was to impart bits of scholarly instructions to them as to how they might accomplish our task.

This is a testimony that airmen can be enterprising and creative when the reward is avoiding labor.

Strange as it may seem, no matter how hard we tried and even with others doing our jobs we still managed to become conspicuously soiled with layers of grease. Our uniforms soaked with pounds of pungent tallow to the point they might self-combust. On my last day of this garbage experience and at my odoriferous peak, I was summoned to the Squadron Commanders office. The messenger added that I was to report as soon as possible or ASAP. Fleetingly, I pondered a side trip to the barracks to render myself more presentable but passed on the idea due to the urgency of the message relayed. I crossed the threshold of the squadron office accompanied with the fragrance of the rancid oil slick emanating from my clothes and my person. As might have been predicted, I was greeted most unkindly. Moreover, it was hastily and pointedly requested that I leave. I tried to explain my presence but was summarily ousted. To this day I have no idea why I was sent for, nor did I ever hear from the Squadron Office again. Perhaps there was an upside to being on garbage detail.

John Childers writes about his favorite rope story:

Some of the ropes did not understand that their authority only extended to the trainees and then only in their area. Some ropes became exuberant in their duties and felt they could order people around. Not so. At the time, I was a Technical Sergeant, a five striper, just reporting in to KTTC for some additional training.

It was raining cats and dogs and I was wearing my Air Force issued blue nylon raincoat which has no outward insignia of rank. As I walked into the BX I heard a loud voice say 'Hey Airman.' Knowing he wasn't talking to me I kept on walking. Now the voice became much louder so I stopped, turned around and I'm looking at an Airman third class wearing his field jacket with his one chevron on his sleeves and a yellow rope around his arm and shoulder. I politely said, 'Are you talking to me?'

His response was to point to the bottom button of my raincoat which was unbuttoned and then to the yellow rope, He then said to me, 'I'll have you know I'm an acting technical sergeant.'

I finished unbuttoning my raincoat and started to remove it. As I did I asked him if he had 'ever seen a real one?' I thought he was going to die on the spot but I gave him some advice. I told him to forget this incident, but to remember his authority was to be used with discretion. Then we parted company, he with a red face and me with my raincoat in hand.

William Scott tells another of a rope:

A rope was taking a shortcut through another squadron's area and began chewing out an airman who was being punished for some perceived infraction. The airman was required to rake the ever-present pine needles. Immediately the rope was surrounded by the squadron members who in no uncertain terms told him if they ever saw him in their area again they would hang him with his own rope. They never saw him again.

Jim Sanders tells this story:

While I was at KTTC, there was a recruit who was a hypnotist. The yellow rope agreed to be hypnotized. Under the spell he was able to repeat the class from the day before as if he were the instructor. Then he was given a post- hypnotic suggestion to make it easier to put him under in the future. Several days later, the hypnotist called the yellow rope over and placed him under again. He was then told to go and kiss the white rope. He approached the man wearing the white rope and tried to kiss him and the white rope ran away with the yellow rope chasing him. Lots of laughing by the airmen, but then the hypnotist realized there was no stopping the yellow roped airman. That was when the hypnotist had to run down the yellow rope to get him to stop. He finally caught up to him and brought him out of the trance. When told of the event, both ropes were embarrassed and very angry but after hearing the rest of the story, the white rope being chased by the yellow rope and both being chased by the hypnotist, they all had a good laugh.

Michael Gordon on messing with the officer corps:

I am reminded of an incident at Keesler. A friend of mine - a military brat - and I were walking over near the BX. This guy had a real dislike for Second Lieutenants (2nd Lts). As we looked down the street, lo and behold, a young 2nd Lt. was coming down the street on a bicycle and carrying a bag of groceries in his right arm. When we got closer, my friend suddenly popped the young officer a sharp salute. Rather than simply acknowledging the salute, the Lt. attempted to shift the groceries to his left arm and return the salute with his right arm. The result was inevitable - the bike slowly listed to the left and fell down. While I was a little embarrassed for the young officer, it was funny. We, of course, kept on walking. If an officer is carrying something and cannot return a salute, he is only obliged to give "At Ease," or return a verbal greeting or acknowledgement.

Frank Kolb, spent four years enlisted service and returned to civilian life:

During my tenure at Keesler, Jan thru Dec '61, Saturday open ranks inspections were held every week and often were followed by a flight line parade. The inspections were usually conducted by junior Second lieutenants who were students in one of the officer Communication and Electronic schools on the base. Many of the members of my squadron, the 3394th, would go out Friday evening and deliberately eat meals heavy in garlic, onion and anything else with strong odors. Indulging in beer was also considered appropriate. Foregoing the brushing of teeth in preparation of the Saturday morning inspection we lined up. Many of the inspectors, no doubt imitating inspectors they, themselves, had suffered under, would make a big production of getting right in your face, eye to eye, with just a matter of inches of separation. I guess it was supposed to be impressive or intimidating or something. After doing the eye to eye thing with a dozen or so of us who were mouth breathing with goat breath, the rest of the inspection was done at a faster pace and without the eye to eye thing. They probably thought enlisted troops were disgusting, we just thought it was fun.

Michael Murphy on making coffee:

One night at Tech school (1972) I had to pull what was called Charge of Quarters (CQ) duty. There had to be someone awake in every orderly room to answer calls and relay messages. There was a NCO with me who asked me (ordered) to make coffee. I told him I never made coffee before as I didn't drink it, He said its not that hard, fill water to the line, put the grounds in here plug in and let it perk. I filled the bottom with water and filled the top strainer completely full with coffee grounds and I mean totally full, plugged it in told him when it was done. He told me to pour him a cup so I did and brought it to him. He took a sip. I thought his eyes would bug out of his head, he gasped that it was a bit strong and went into the latrine. He never asked me to make another pot and strangely enough, I never had CQ duty again.

Jim Sanders tells of a blanket party:

We had a guy at Keesler who, I swear, looked like 'Zero' in the Beetle Bailey cartoons. He wouldn't take a shower and smelled like a wet Airedale. He bragged about still having his first bar of soap from basic training. He was reported to the rope and they were ordered to do a blanket party on him. They first roughed him up under the blanket a little and then hauled him to the latrine for a shower. Some of the guys were gagging from the smell, but they scrubbed him down pretty good using his bar of soap and some stiff bristled brushes. He got the message, but would ask one of us every day if he needed to shower. He couldn't understand to take a shower daily on his own. He would also forget to shave. The NCOIC made him dry shave in front of the squadron one morning. His face was cut all over. I felt sorry for the guy. I don't know how he made it through Basic at Lackland.

Bob Irving continued with his experience:

During my first week at KTTC I was assigned KP. On the second week we did nothing but run and march and perform physical training. Finally on the third week I started my technical training but still had another week of basic to do so we were up early in the morning to get ready for an open ranks inspection then march until it was time to go to school. We were taught from 3:00 until 9:00 pm then marched to chow and finally were able to get back to our rooms and drop dead. Friday night it was spit shine the floors and shoes, iron our uniforms with enough starch you had to get on chair to get into them so as not to break the crease.

Bob Fogarty related this about USAF assignments

Before graduation from KTTC, we all had to fill out what were called "Dream Sheets". This was a form where each man had to list his preferences for assignments. We had to list three overseas locations and one stateside assignment. These were to be sent to Air Force Personnel Assignments and were to be used to try to place people where they wanted to be. For some reason it seemed that no one was ever assigned to where they desired. It was rumored the forms were sent to the "Practical Joke" department run by the division of "Fairy Godmothers" Further the rumor went on to describe that the department only had one employee, an old civil service government service level three, who if she was not out on sick leave would occasionally put down her knitting and select a dream sheet from the stacks that surrounded her and that person would get the worst of his selection. All others graduating from KTTC went to where they were needed no matter where they wanted. Usually they were needed at chilly foreign lands like Greenland or Labrador. They might get other chilly places like North Dakota, Minnesota, or Alaska. They would only be assigned to a place like "outer-slobovia" far removed from civilization. Knowing this we filled out our dream sheets with skepticism.

I was fortunate in that I would be assigned to a base with aircraft as my training had been in aircraft radar. The guys I felt sorry for were the ground radar guys. They were going to be assigned to sites and being as they were new young airmen they were almost guaranteed an assignment to some remote mountaintop about a zillion miles in the boondocks, where the mercury was frozen at the bottom of the thermometer year round. It would be to a place which could not be located on a map with a name that even locals could not pronounce. A place where the mail would be brought in by dog sled every few months and they might go years without seeing a girl. Even worse would be that the airman's club would run out of beer or that it was on the verge of running out and the beer would be rationed, a half a glass a day per person. Perhaps it would be on some Aleutian island where in your off duty time you could go to the north side of the island and pick up big stones then walk the half mile to the south side and drop the stone.

Eventually the island would be moved closer to the states. The tests for when a person had been out there too long was when they would start cutting out paper dolls, flirting with the local dog sled dogs or wearing a straight jacket as the uniform of the day. All so they could detect the hordes of evil communists bombers expected to attack the United States. God bless them all.

THE PEACEKEEPERS

CHAPTER EIGHT

Getting to their sites

When training was completed at KTTC, the men were upgraded from untrained to apprentice and assigned to a site. On arrival they most often were used to test the security alertness of the other airmen assigned. They were also scorned as being the newest member of the team and taunted into embarrassing situations. This was especially true at remote and isolated sites where they might have to be there for the next thirteen months. Everyone else had less time to serve before they went back to the "world". There were often initiations all in good taste but at the expense of the "newby". This was the time that the newby had to prove themselves to the others.

Jim Sanders was a Sergeant and a member of AFRSV asked this:

How many of you were used as a guinea pig to test site security when you first arrived at your radar site? I had on my blues and no nametag yet. The security officer and one of the Air Police informed me of what to do. I was to walk into the end door of the operations building and at the end of the corridor I would find a locked door with a pushbutton lock. I was to wait until someone came out the door and I was to barge into the scope area. Well I was hanging around the door when someone came out. They didn't recognize me and saw no badge. They began yelling "Seven High, Seven High, Seven High!"

That was the alert code for we are under attack. Then all hell broke loose. I was pushed against the wall. Two guys with rifles escorted me out the side door and held me at bay. Then an air police van from main base showed up. Out piled several skycops with full banana clips in their carbines and pointed them at me. I was slammed against the side of the van. Ordered the spread my legs and they frisked me. They checked my pockets and even the band of my hat. After intensive questioning the security officer showed up and terminated the drill. I was turned loose and was still scared out of my wits. I was happy I hadn't been shot. The guys and I later had a big laugh about everything, after we knew who each other was and what we were doing there. Some time after that I spotted another new arrival being prepped to try the same thing I went through. I caught him half way down the hall and the Air Policeman called the thing off, 'cause I had ruined it for them. I spared that poor airman the same ordeal I had gone through.

George Culp wrote of his first duty station at Point Arena, California:

There were about 30 new airmen in the radar maintenance section. One of the three stripers sent a new airman to supply with a written requisition for a fallopian tube telling him it was needed to fix a piece of equipment in the radar. The new airman was sent to Supply to get this exotic tube. After laughing like crazy, the supply Sergeant asked the airman if he knew why he was laughing and if he knew what a fallopian tube was. After the airman admitted he didn't, the supply Sgt explained it to him. The new airman stayed cool and asked the supply Sgt to write "Not in Stock, On Backorder" on the requisition form. He took it back to the three-striper and handed it to him, and said with a straight face "Supply said they were on backorder." After a minute or so of puzzlement, followed by a lot of jabbering on everybody's part, everybody broke out laughing. The rest of us airmen were amazed, and proud, that the airman had turned the joke around so nicely. I was glad it wasn't me who got tricked, and I'm certain the others felt the same. I don't think any of us knew what a fallopian tube was either.

Paul Goldschmidt was at Baudette, Minnesota:

We often sent the guys right out of tech school to supply, either for fallopian tubes or radar paints.

David E. Casteel As an Officer, I would never do anything like this myself, however:

Our new guys were sent for Angle Markers, Radar Paint, Cable stretchers.... What else?

Kerry Lushbaugh, a radar site vet:

How about Sky-Hooks and Video Paint?.

Rob Plested:

Wrench benders?

Robert Chaney:

We sent them out for Right-handed smoke shifter.

Bob Fogarty:

Probably the best was the guy that couldn't understand a German radio broadcast, and got sent out looking for a replacement translator tube...

Robert Jordan:

A bucket of RF was always good for a laugh.

Jim Sanders:

Anyone recall the "Telco Sensitivity check"? You'd call the scope dope on the phone and ask if he had a "zero beat oscillator" to do a sensitivity check? When he didn't you'd ask him to step back and whistle into the phone? It worked most of the time.

Tom Terrell CMSGT Retired:

The sensitivity check sounds like the old CANNON report call. We would call some outfit if we knew they had a "newbe", someone new and inexperienced. We would ask whoever answered to put the new guy on and tell him to "Stand by for cannon report." There would be a pause then using the phonetic alphabet we would give him, "BRAVO, OSCAR, OSCAR MIKE. Stand by for disregard." He wouldn't know what we were talking about and the questions would come from the newbe to the others who would string him on. Ten minutes we would call back and give him a "Disregard." For those ten minutes he would be standing there with a dead telephone at his ear and a confused look on his face.

Gaston Cannon, Jr.:

I hadn't been at the RADAR site at Lackland too long, and was behind the plotting board one night when I got the Cannon report: I guess that's also when I got the nickname of "Boom Boom".

Bob Wildrick:

I don't remember the "Telco sensitivity check" but I do remember one scope dope being sent around the 902nd trylng to get some dial tone. He had asked the comm guy where dial tone came from and was told it is the clear liquid in the batteries. Was also great stuff for car batteries, made them last longer. I know he was sent down to the radio shack as I was on duty and sent him to motor pool and from there I don't know where he went.

Bud Bussell was in from 1955 to 1959 and served on AC & W sites:

One midnight shift some "Scope Dope" (A1C type I found out later), I was A3C, came down the hall hollering that there was something wrong with the scope on the upper right dais. I guess I was the one set up, newbie me, as I was the first to respond. I ran into that black environment, no time to allow eyes to adjust, and grabbed the steel ladder and started up. OUCH.

Someone had taped about two or three thumbtacks onto one of the rungs of the ladder. I got two in one hand; one went fairly deep and really stuck into my palm. When I heard the laughter, I let everyone know that I was not a happy camper. Not sure if it was because of the tack in my hand or my embarrassment at not seeing the tacks.

Later that night the Officer in Charge of the shift came into the maintenance shop and offered his apology for the incident. I never said a thing about it after that and never heard of it happening again.

Joe Sergent:

Way up there in Cut Bank Montana, back around Brown Shoe Corps days, they used to send Newbie Radar Techs and Operators all over the Squadron for either " Prop Wash " "Range or Azimuth Markers " " Fallopian Tubes" " IFF Paints " etc. This was usually a long drawn out affair as the Newbie would first go to the Parts Stock Supply who would send them to the Radio Transmitter Towers usually up the hill at one end of the base, then to the Radio Receiving Towers at the other end. Then they would be sent to the Civilian Engineering Shop, Motor Pool and back to us with either a "No-Go" or some made up facsimile. One guy came back with a small can of black paint for the " Range Markers " and was told jokingly to put them on the radar console and actually started to paint the Range rings on the scope before someone saw he was really starting to do just that and stopped him.

Bob Edwards, Technical Sergeant:

Talking about practical jokes, we had this new A1C show up, only place he had ever been was at some site in California, and all he ever had to do was be available to push the red button. First day he was on my crew, I sent him down to supply for "search paint." I called the supply man to warn him. He sent him back with nothing saying they were out. So I sent him back for "SIFF paint as a substitute. Well, he was gone about 30 minutes finally came back saying supply was out of it also. This was a lie, because Charlie in supply said he never showed. I chewed on his butt for about 15 minutes, everyone else in Operations was laughing so hard I couldn't keep a straight face either. I finally let him know about the joke.

Robert Wilder:

When I was a Private First Class with the New York Air National Guard, I was called to active duty and assigned to Grenier AFB, New Hampshire; I ran into the joke and did my best to reverse it. I was instructed to get a bucket of "prop wash: and a "sky hook." I spent about two hours and found the items. The "prop wash" was easy. I got a bottle of detergent and wrote on the bottle "special detergent for washing wood Hamilton variable pitch propellers" used on some 1940 aircraft. The "sky hook" took a while longer, but having the NCOIC of the motor pool being from my home- town, I delivered the "sky hook" to the First Sergeant. This is a large steel ball with a hook through its center used on a crane. I had the Motor Pool drop it on the floor in front of the First Sergeant's desk. Needless to say I was never asked to go for any other foolish items.

David Elkins:

There was none of that ridiculous sending a new man all over the site to get something that did not exist at the 914th Radar Squadron, Armstrong Ontario Canada. When a new man arrived he was the guest of the club for the night. He was given a drink and he was welcomed. The drink was a shot of everything behind the bar with enough crème de menthe to color it green. After all we were part of the pinetree line. The new guy had as long as eight hours to finish it if he could. Many men tried but passed out before it was gone and some men declined for personal reasons with no hard feelings. It was a good night to rub it into the new man that he had more time to go than anyone in the club.

Jay C Phillips who made the USAF a career and retired with the highest enlisted grade possible, Chief Master Sergeant, tells this:

My assignment to Killeen AFS Texas was the most unique radar maintenance assignment I ever encountered. It was July 1958 and I was a Staff Sergeant with four years in service. I had just graduated the previous month from Keesler AFB, and had been assigned as a radar maintenance apprentice. My graduation orders sent me to James Connally AFB, Waco, Texas.

On 3 July 1958 I located my new squadron and the First Sergeant informed me I was being reassigned Permanent Change of Station with out Permanent Change of Accountability. He handed me a new set of orders, assigning me to Operating Location Detachment 1, 3568th Navigator Training Squadron (Flying Support), Air Training Command, Gray AFB, Killeen, Texas. He then told me to report to my new assignment.

West of Killeen, Texas I found a small sign proclaiming Gray AFB and turned onto a narrow macadam road. Arriving at a guard post, I had my ID and the order closely examined. Then I was told to continue on and not to stop my vehicle till I reached the next guard post. I was getting nervous with all this security.

I had gone about four miles when I came over a small rise and there was Gray AFB, and then the adventure began! At that time it was all highly classified. Every one assigned to the installation was provided a special identification badge that each person carried and was strictly responsible for. Next to Gray AFB was Killeen Base, a Defense Atomic Support Agency manned by US

Army Military Police. Built into a series of hillsides were special protective bunkers for storage of nuclear weapons. Killeen Base was a large area of small hill's and was thickly wooded. The perimeter of the special weapons storage area was surrounded by an electric fence, which was not unlike what you see in movies or television shows. First sterile gravel covered the ground, and then an eight-foot tall chain link fence topped with coils of barbed wire. A ten-foot wide asphalt roadway separated the outside fence from a second inside chain link fence, which was electrified. All the electrified fence posts were mounted atop large insulators. Inside this fence was another roadway on which heavily armed three man patrols in jeeps (equipped with a pedestal mounted .50 caliber M2 machine gun) roving back and forth on the perimeter roadway as well as the inside roads. Needless to say the security at this place was tight!

Gray AFB had one runway, a small control tower/operations building and a minimum of base facilities. Its primary function was to support the aircraft that transited the base carrying special weapons.

The 814th AC & W Squadron administrative and housing function for single airmen was located in a new two story building. That's where I found the office for Detachment One. Two other newly graduated maintenance men and I were assigned to maintain six radar consoles for the 12 or so detachment officer's who were assigned to provide navigational and intercept training for student officers assigned to James Connally AFB. The six consoles were located on top of a three level dais. On paper this looked OK, but in reality three of us could not obtain our upgraded skill level from apprentice to skilled without becoming knowledgeable on all of the station's radar equipment. We requested and were approved to be integrated into the unit's maintenance work schedule. We were now ADC airmen assigned to an Air Training Command unit, located on an Air Material Command Base and, drawing part of our support from the US Army. Fort Hood, just north of the base, was a sprawling 340 square mile installation that supplied our food, pay, and medical support plus creature comforts such as the Post Exchange and Commissary.

In addition to everything else we found the station wasn't even operational. There were no operation's personnel, just a few radar maintenance technicians. In those early days there was only one man assigned to the teletype section in the communications room which was on duty 24 hours per day. So along with other duties, radar maintenance technicians monitored the teletype machines during night time and weekend hours. Our equipment consisted of a Search Radar, a Height Finder Radar and a dark room full of radar consoles. There was also the requisite video mapper and target simulator equipment. The radar antennas were on temporary towers. The operations area was reached by driving out the back gate of Gray AFB through another guard post, four or five miles and, then up a very steep road to the top of a flat bluff. There the only buildings were the operations building, supply building, the diesel power generating building and a small Ground to Air Transmitter and Receiver (GATR) building. While we were getting the bugs out of the new radars and associated equipment, enough operators became available to go on line for eight hours a day, and a month later 16 hours, and in the following month around the clock operation. In November steel beams and other construction materials arrived and construction commenced for permanent radar antenna towers. We were told that funding for more new radars had not been approved but could get funds to modify existing ones. During this construction we were required to cease operations so we wouldn't radiate the construction crews. Most of the maintenance people were sent on temporary duty status to operational sites for training on the new radar we would soon be responsible for. I and two others ended up at the 703rd AC&W Squadron, Texarkana Air Force Station, Texarkana, Arkansas.

Interspersed with our normal work load were other interesting events; Such as the time an operator swore he had seen a rattlesnake crawling out of a cable trough behind a radar console. The "Great Snake Hunt" was on. All of the cable trough covers from the towers to and, into, the operations room were removed in a futile hunt for the phantom snake. To deter any future possible unwelcome guests all cable trough entry points in to the building were sealed using copious amounts of screen wire and plaster-of-paris.

Our work uniform was a one piece cover-all with belt; very hot and hard to get on and off. It was not one of the better uniforms the Air Force issued. However, with the Army nearby the one piece cover-all style fatigue uniform was quickly supplanted with Army two piece fatigues as that is what was available for replacement. Remember the Army was our supply provided so when a pair of those one- pieces were ruined with grease or were torn, and we made certain they did get greasy and torn and were quickly exchanged for the Army's two piece fatigues.

William Decker:

While stationed at 716th AC&W in Great Falls Montana, I had occasion to see one airman new and green as they come given the duty to be a security watch. We were on a mountain 7300 ft above sea level and having a blizzard. This dumb airman was sent outside for ' hurricane watch." I went looking for him awhile later and found him standing outside in the doorway still watching. I talked to him about his duties then finally clued him in that hurricanes do not occur in Montana, especially in a snow blizzard.

Wayne Fitzgerald, also a retired Chief Master Sergeant:

Speaking of desolation, I have a question to ask the members. How many of you were ever stationed in places that had no barber/s and had choose up sides and cut each other's hair until the best ones surfaced and was given the job as site barbers? That's how we had to do it when the existing barber returned to the world.

William Lechien:

Tatalina Alaska had a tradition when I arrived there in Dec 66. They had a very large rope tied into a noose which was passed to the newest arrival with a little ceremony. When I arrived, the man in possession of the noose, met me getting off the aircraft, placed the noose around my neck. He placed his arm around my shoulders and whispered while pointing to all the men there "Every one of these guys are gonna leave here before you". I went along with their ceremony, knowing that what he said was not true. They didn't know my situation.

I was on site about a month when my name went up on the board in the NCO Club as being one of the ten shortest men on site. It was only then that I revealed that I had spent eight months on Fire Island and was transferred to Tatalina to replace the computer tech. that had been sent back to the lower 48 with a bleeding ulcer. I inherited his date for return from overseas (DEROS, pronounced De-ros) and only spent 11 months in Alaska as a result. Everyone had a good laugh and I passed the noose to the latest arrival without the ceremony.

Vic Tanner:

I didn't get the noose when I arrived at Tatalina, but on arrival at hilltop I did go through an initiation ceremony. You were told that you had to drink a beer while standing on your head. They poured you a glass of beer, turned you on your head and placed the beer on the floor in front of you. You were supposed to pick up the beer and drink it while inverted. What you didn't see was the pitcher of beer they had hiding around the door next to where you were inverted. When you went to pick up the beer someone would reach around the corner and pour a pitcher of beer down your pants legs. What a rush. We all had a good laugh and of course when I got to be the short timer I was the lucky one doing the pouring. It was my honor to get the new hilltop NCOIC going thru the initiation.

Bill Porter:

I remember performing routine maintenance on the FPS-7 standby channel. We posted warning signs not to turn on the standby channel and went to work. Next thing I knew I heard the loudest bang that I have ever heard! The guy I was working with was supposed to discharge the high voltage capacitors on the standby channel, but had put the grounding rod on the live channel capacitor.

127

The tip of the grounding rod was melted and he was on the floor in shock. We rushed him to the medic and he was declared OK. I seem to think the interlocks should have prevented this. He was careful after that. It was a hard way to learn.

Larry Cameron tells of his nickname:

I got a nickname when I entered the Air Force. I was known as Slick Cameron even though my real name was Larry. I always answered to both names and still do.

This was in the days when we were paid once a month and had to get all dressed up into a class A uniform, shined shoes and necktie, the whole shebang. Then we had to get into a line in alphabetical order and when it was our turn to receive our pay we had to approach the pay table and the paymaster who was an officer, salute, and state our name and that we were reporting for pay, while holding the saluting. He would return the salute and go through our squadron index cards to determine if there was any reason not to pay us, in other words, had we been fined for some infraction. If the index card was clear he would locate our name on the pay roster, count out the amount of money we were to receive and have us sign the pay roster. My turn was up so in my finest manner I stepped to the table and saluted, stating "Sir, Airman First Class Cameron, Larry, reporting for pay." The paymaster was a Second Lieutenant who knew who I was but after I reported he got a perplexed look on his face as he went through the index card file looking for Larry Cameron. Moments went by and finally he looked up and said "there is no Airman First Class Cameron here." I was caught off guard and stepped backwards just knowing I was going to have to do battle with finance to get some money. Then I realized, the card file was from the squadron so I stepped forward again and saluted this time saying "Sir, Airman First Class Slick Cameron reporting for pay." He again thumbed the card file, smiling as he pulled a card and said, "Sign here Slick," and he paid me. This got to be a fun ritual every month until I rotated out of Cartwright, Labrador.

Don Williams:

I was stationed at Fort Yukon, Alaska in 1968-'69 in radar maintenance. There was a Civilian who lived in town who was a pilot with his own plane which he parked in his front yard. The plane was a small two-seater and he used several cinder blocks as tie-downs. These were attached at the end of the wings on his plane by ropes. He also had a little white poodle which he dearly loved and took most everywhere.

Anyone that knew this man also knew he liked to drink. One day he had too many suds and the urge to fly was greater than his intelligence. He and his little poodle went for a flight. Besides the obvious problem of a drunk flying an airplane, when he untied the tie downs, he only untied one side. He managed to get the plane air born even with the cinder blocks hanging from the left wing. In the process of taking off, he busted up most of the cinder blocks but a couple was still intact and was really weighing down that side of the plane. So much so that he couldn't get any altitude and he couldn't do anything but fly in a big circle just above the trees and with that wing tilted down. He managed to get out to the site and make a landing onto the old dirt runway. It busted up his plane as the blocks were bouncing on the ground and hitting the wing. He got out with a few cuts and burses and the little poodle was not hurt.

It was a long time before he would or could fly the plane which was probably a good thing because it did not cut down on his drinking.

John Couch:

Chief Master Sergeant JUST D. ROCHE was the First Sergeant at the 679th AC&W Squadron located on Tyndall AFB, Florida for 11 years, from 1960 to November 1971. He had served in WW II. He was highly respected by the troops and really went to bat for them. I remember one promotion cycle when he was a member on the 14th Air Force promotion board. There were 14 Staff Sergeant Stripes allotted to the 32 Air Division and the 679th got 13 of them. Major General Putnam, Commander of the 14th Air Force called Chief Roche directly after the promotions were released and told him that he would never sit on another promotion board as long as Putnam commanded the 14th AF. Did not matter to the Chief, he got the stripes for his troops.

Later General Putnam called Chief Roche and told him to submit his retirement papers that he did not want a First Sergeant in his command who was the First Sergeant of a squadron that had totally failed an IG inspection with major problems in personnel management and the condition of the barracks. The only problem was that the squadron that failed was the 678th not the 679th. Later General Putnam apologized to Chief Roche through the squadron commander not directly to the Chief.

Pete Holland:

I was a little dismayed to find that we had to stand on open locker, standby inspection for a visiting Navy admiral who arrived on site to do a little goose hunting.

As we stood at attention, the admiral went through the barracks, wearing his hunting jacket and waders. In my room, the admiral asked "Airman, why do you have a two-foot boulder in your room. (Some of site-mates had placed the boulder in my bunk on one particularly drunken night.) My reply to the admiral was: "He lives here, sir." The admiral just went down the hall shaking his head.

Wayne Fitzgerald:

I also have a rock story, and I still have the rock. I was with the 102nd Tactical Control Squadron, Rhode Island Air National Guard. In 1976 we were ordered to spend the winter at Tonapah, NV. We had our site at Booker Mountain east of the town and we had our usual problems with proper groundings for the equipment. Airmen were trying to dig a ground pit by hand but because of the number of rocks, they wanted to rent a back-hoe to complete the job. I didn't have the funds for such a rental and tried to show them that the rocks weren't that big. They never forgot. Our commander got the funding for the rental, the pits finished and the equipment grounded. I didn't hear anymore about it until we got back to Rhode Island. I was leaving the unit and we had a quick good bye. On the table with an object covered with a sheet. As we were wrapping up the good-byes, the airmen called me over and read me a poem about my lack of understanding what a "rock" was. They gave me a rock that weighted about 70 pounds from Tonopah. They had it shipped it back by C-141 to Westover with our other equipment and brought it down to the site to give to me.

I have never forgotten the unit, and as I moved on to other assignments I have carried it with me. It is still with me in Tucson, Arizona and has a place of honor on my cactus mound. My only problem is that it would like to go back to Tonopah for a visit. It doesn't know if it would like to live there, but a visit would be nice.

William Smith:

In 1962 as an Airman Second Class I was on duty in the GATR building dressed in my fatigue uniform. My NCOIC told me I needed to report to the Squadron Commander in dress blues in 30 minutes. It was five miles to the barracks so I was rushed for time to make the trip and change uniforms, my blues were still in the cover after just coming back from the cleaners. I changed and made it to the site on time and reported to the Commander as directed. After he returned my salute he looked me up and down and said "Airman Smith do you realize you do not have any collar insignia on?" Then, before I really thought I said, "No Sir I didn't, but do you realize you don't either?" He looked down at his collar and said, "I suppose we both need to go put some on." Until this day I still do not know why I was to report to him. I asked my NCOIC and he didn't know. I am not sure of what kind of impression I left on the Commander, but I did get promoted to Airman First Class with minimum time in grade.

Walt Bussell, wrote:

Aw you just hit the memory bank dead center. I was stationed at the 799th AC&W, Joelton AFS, Tennessee for six months. October '56 to March '57. Lot's of heavy work as the station was not operational yet. We unloaded equipment from railroad flatcars and used railroad ties to line the road inside the fence. We had no forklifts, just muscles. I volunteered to drive a truck because I was

145 pounds fully clothed and soaking wet. I drove the ration truck three times a week to pick up the food in Murfreesboro. Up early and back around noon. Then I had to finish the day in Radar Maintenance. KP was terrible there my name had never come up on the rooster. The only people excused from KP were the Air Police. Someone asked me why I did not pull KP. I said it was because I was the truck driver for the mess hall. That turned out to be the wrong answer and caused me many hours of grief. It was my name that got me in trouble. Seems there was an Air Policeman with the name Fussell, a nice kid but the only sky cop to be assigned KP. Somehow Bussell, not Fussell, had been scratched from the KP roster. They got it corrected by assigning me to five straight days of KP to get caught up. Ugh.

George Holmes:

I guess unless someone has been to both it would be hard to determine which was worst. I was at Sparrevohn Alaska, 68-69 at their top camp. As I understand it Indian Mountain and Sparrevohn were the only two with "full top camps." At Sparrevohn we had about 60-70 people and bottom camp had about 100 plus or minus. I assume Indian Mountain was much the same. Depending on your point of view a smaller or larger group of people to associate with is either good or bad. Obviously, the top camps didn't have as much to offer as the larger bottom camps at other sites. In the summer I hiked to the fish camp and wandered around. I rode and walked up and down the mountain several times, but probably didn't spend more than a dozen hours at bottom camp.

I visited the guys at the weather station a few times. However, after saying all that, I'm not sure I would have

133

liked being on a "large" site where three or four radar maintenance dudes had to rotate to the top of the mountain for a week at a time. Regardless, it was a life experience that still provides stories today.

Charlie Leach:

My start in the Air Force was like many others I'm certain. I had no money to attend college so I enlisted in August of 1957 I was sent to Lackland for a real change of life. After a couple of days I was asking myself if this was what I really wanted. I got with the program, learning how to clean the baracks with a toothbrush, spit shine my shoes, stand in long lines and wait, and the finer points of life on being an Airman. After the fourth week of phase 1 basic training, I was sent to Keesler for AC&W School. Wow; basic two hours a day and school six hours a day, on second shift. Three man rooms certainly weren't like the thirty man open bay barracks at Lackland. I didn't have KP at Lackland but did at Keesler. School was easy, finished as class honor graduate. Next came assignments. Mine was to Boron AFS, California. Where was Boron? As luck would have it, one of my instructors had been stationed there. He told me, you take a big airplane to Los Angeles, a little bus east to Boron, and then call the base for transport service. I did that and called the base upon arrival. About 30 minutes later, an Air Force car arrived filled with three guys. The driver took the back road out of town; seems like there was a problem with the airman from the base and some of the local young ladies of the fair city, so the exit from town was quick. I was assigned a room for the night. The next morning I looked out the window in every direction for 50 miles and saw nothing but sand.

It was just before Christmas '57 and I was offered KP

over Christmas or New Years. Boron had a KP policy, 10 days of KP about once a year, followed by a three day pass. This year the Commander was giving seven day basket leaves also. A basket leave is one where all the paperwork is filled out and not turned in so it is not charged against you. However, if you get in trouble or don't get back on time, then the leave request is submitted. I took Christmas KP, then, had 10 days off.

In mid 1959 I became a member of the premier crew which worked only day shift and on occasions other shifts during special missions. Generally I was on the day shift and we had week-ends off. I had a '57 Mercury convertible and my room-mates parents lived in Hollywood California. I had the summer of '60 planned. Drive to Hollywood every weekend and spend them on the beach. Life was going to be good. Well, let me change that. Have you ever heard of Murphy's Law? It goes like this. "If everything is going well, you obviously have overlooked something." Sometime it is, "If anything can go wrong, it will and at the most inopportune time."

I was informed in April '60 that my GI Butt had been requested by the 10th Air Division in Alaska. I guess I would not be spending my summer at the beach after all. The night before my flight to Elmendorf AFB, AK, I shared a room with an airman just returning from there. He gave me a clock, it wasn't much, but it did have an alarm. I arrived in Alaska about 2100 local and was assigned a bunk in the temporary duty barracks. I set the alarm for 0600.

The alarm went off and I looked at my watch it was four o'clock. The sun was bright and there I was on my first day and I'm really late. I dressed quickly and started to leave when I realized I was the only one up, everyone else is still in the bunk. The sun shines bright at night in the summer and the real time was 0400. I threw the clock into the trash and went back to bed.

During the week of preparation for Alaska living, my group was asked if there were any ham radio operators in it. I was a ham operator and in fact brought my radio to Alaska with me. That was my cue. I was given my choice of assignments. My new home would be with the 626th AC&W Squadron at Fire Island AFS, Alaska.

There were no hams on Fire Island. Since I was the new kid-on-the-block I didn't get into the maintenance and inspections section, I just became a scope operator. Within two weeks I had been authorized to operate the Military Auxiliary Radio System (MARS) station in the afternoon and work ops in the morning, day shift. Within a couple of months I was the full time MARS Station operator and no more scope watching As the ham assigned full time to the MARS Station I had a great job, MARS nets daily, I ran the High Frequency net for Alaska, phone patches at other times, and even got promoted to A/1C. I even served as a single point communication system for backup to White Alice, which connected remote Air Force sites in Alaska to command and control facilities. In some cases it was used for civilian phone calls. After the Air Force, I worked in Broadcast Engineering in top TV Stations, completed college and retired as an instrumentation engineer with one of the government contractors. I had a great time. The Air Force active duty was the high point, the rest was just gravy.

Issac McLemore also known as "Walking Hawk". He would not say why:

When I was at Topsham AFS, England I was primarily in maintenance but also helped out the central computer and display guys. On mid-shift, we began to get calls from the scope guys telling us the picture on a particular screen wasn't clear enough. This was a random malfunction, not limited to any particular scope or day. We would check certain adjustments, maybe tweak some a little, and clean the screen like we did on every scope call. Although we couldn't say exactly what we did, this would always clear the problem.

After about a month, we figured it out. Remember, this was in the middle 60s, when some guys liked to sport long hair when off duty. This one particular scope guy's hair fell well below his collar when off duty. He was on permanent swings, and when he reported for work, he would have his hair greased down flat. Sometimes late in the shift, he would take a quick nap with his head on the scope table. Of course, his hair would brush the screen and off came the grease. Then the guy following him on the scope on mids would call us because his picture was bad.

Once we had it figured out, we would send any new guys on the calls and watch them scratch their heads trying to figure it out.

An unusual side effect of radar was identified by local police units. In later years, Officers reported that when running speed traps using their radar guns, the speed indicators would give false readings. It was suspected that when the radar antenna revolved and went over an area where police radar was running the USAF radar would interfere with the police radar units. Radar site commanders at sites located where complaints emanated warned that if any airman intentionally aimed any radar mast towards a highway, especially a highway used by airmen returning to the sites in order to cause false readings by police radar, disciplinary action would have to be taken.

THE PEACEKEEPERS

CHAPTER NINE

Handling their off-duty time

Some radar sites had swimming pools disguised as water ponds required for fighting fires. Some commanders even had to issue orders not allowing swimming in these water retention basins. Some sites had camps where the personnel could go for up to three days. Camps were in the middle of nowhere, usually next to some remote lake. There would be a boat stored at the camp and the personnel would go to special services and check out the keys and a boat motor. Several sites had power boats with big motors and water skis. Most all had theaters, pool tables, ping-pong tables, hobby shops for ceramics and leather work, and a library routinely stocked with books and magazines brought to the site by the personnel themselves. Those sites located near civilization had access to television and radio. The remote sites had their own radio stations with the volunteer disc jockeys being off duty airmen.

There were several possibilities for some of the airmen to be employed in jobs such as bartender at the clubs, barber, Base Exchange store manager, laundry, pin setter for bowling lanes if they had any lanes installed and theater manager/projectionist. Many airmen took correspondence courses and furthered their education. Some collected sayings.

Gaye Cline offered these:

"Airspeed, altitude and brains, two are always needed to successfully complete the flight." - Basic Flight Training Manual

"Mankind has a perfect record in aviation - we have never left one up there!" - Unknown Author

"Flying the airplane is more important than radioing your plight to a person on the ground incapable of understanding or doing anything about it." - Emergency Checklist

"The Piper Cub is the safest airplane in the world; it can just barely kill you." - Attributed to Max Stanley (Northrop test pilot)

"There is no reason to fly through a thunderstorm in peacetime." - Sign over Squadron Ops Desk at Davis-Monthan AFB, AZ

"If something hasn't broken on your helicopter, it's about to." - Sign over Carrier Group Operations Desk

"You know that your landing gear is up and locked when it takes full power to taxi to the terminal." - Lead-in Fighter Training Manual

"It is generally inadvisable to eject directly over the area you just bombed." - US Air Force Manual

"Whoever said the pen is mightier than the sword, obviously never encountered automatic weapons." - General MacArthur

"You, you, and you ... Panic. The rest of you, come with me." - U.S. Marine Corps Gunnery Sgt.

"Tracers work both ways." - U.S. Army Ordnance Manual

"Five second fuses only last three seconds." - Infantry Journal

"The three most useless things in aviation are: Fuel in the bowser; Runway behind you; and Air above you." - Basic Flight Training Manual

"Any ship can be a minesweeper, Once." - Maritime Ops Manual

"Never tell the Platoon Sergeant you have nothing to do." - Unknown Marine Recruit

"If you see a bomb technician running, try to keep up with him." - USAF Ammo Troop

"You've never been lost until you've been lost at Mach 3." - Paul F. Crickmore (SR71 test pilot)

"The only time you have too much fuel is when you're on fire." -Unknown Author

"If the wings are traveling faster than the fuselage it has to be a helicopter - and therefore, unsafe." - Fixed Wing Pilot

"When one engine fails on a twin-engine airplane, you always have enough power left to get you to the scene of the crash." - Multi-Engine Training Manual

"Without ammunition, the USAF is just an expensive flying club." - Unknown Author

"If you hear me yell; 'Eject, Eject, Eject!,' the last two will be echos"- Pre-flight Briefing from an F-104 Pilot

"What is the similarity between air traffic controllers and pilots? If a pilot screws up, the pilot dies; but if the Air Traffic Controller screws up, then the pilot dies." - Sign over Control Tower Door

"Never trade luck for skill." - Author Unknown

Robert Jordan collected these:

Air force definition of explosives: A loud noise followed by the sudden going away of what was once there a second ago.

Always remember that your weapons are made by the lowest bidder.

Bravery is being the only one who knows you're afraid.

Cluster bombing from B-52s is very, very accurate the bombs always hit the ground.

Combat will occur on the ground between two adjoining maps.

Don't draw fire. It irritates the people around you.

Don't ever be the first, don't ever be the last and never volunteer.

If it's stupid but works, it isn't stupid.

If you find yourself in a fair fight you didn't plan your mission properly!

Never share a fox hole with anyone braver than you.

Never trust a private with a loaded weapon, or an officer with a map.

Sometimes I think war is God's way of teaching us geography.

The enemy invariably attacks on one of two occasions: 1. When you're ready for them. 2. When you're not ready for them.

The more you sweat in peace, the less you bleed in war. (Old Chinese Saying)

Please flush twice- It a long way to the mess hall.

Man deserves to enjoy the consequences of his own damned foolishness.

Brian A. Coy. I was at Murphy Dome, Alaska in the early '60's as a NCO:

There was a running gag in the NCO barracks. Somehow, a full size fire hydrant ended up not attached to any water line. Although it was not attached to any water line, it was still nicely painted bright red. The practice was to have a room assigned to incoming NCOs ready before they arrived. The floors were cleaned, room dusted, bed made, and the closet and dresser cleaned out. We also set this fire hydrant in the middle of the newbie's room, and explain to him that it was a design mistake, but there it was. The hydrant was heavy, so sometimes it was several days before the occupant of the room realized it was not attached to anything. When I arrived it was in another's room so I was denied the temporary ownership of the hydrant.

Joe Sargent. I was at Cut Bank Montana from '57 through '61 also as a NCO:

We spent our off duty time in town, the NCO club or hanging out in the top floor of our barracks with our Hi Fi's and guitars. My friends and I from made some sleds from sheets of plastic and a tobogan of sorts from a sheet of plywood, 2 x 4s and ski's from Special Services. We used to slide down a hill passing the NCO club. One time we made a night run. Sliding off the hill we knew we were approaching the edge of the butte and knew that it was steep. We were travelling at Olympic quality luge speed. Being trained ADC airmen armed with the knowledge that discretion was the better part of valor, the five of us jumped off the make shift toboggan when we saw the farmer's barbed wire fence speeding towards us in the moonlight. The toboggan continued under the fence and was gone forever. I always wondered what the farmer's reaction was that next spring or summer or whenever he happened upon that strange contraption sitting out in the middle of his field below the radar site butte in the middle of nowhere. He probably thought it had been dropped from the sky by aliens.

Richard Hazelmyer. While stationed in Germany:

One night my television set stopped working. Being as I worked with trained electronics technicians; I took the tubes to the base and got them checked. They all passed the tests. While putting them back in the TV, I looked down and found that the plug had fallen out of the transformer socket. Dumb, dumb but then I did have something to do that night.

And these from the author, **Jack Miller:**

Airman First Class (A1C) Ernie Lake was a character. He loved a good joke even if it was played on him. In addition to his regular duties he was a bartender at the enlisted club at the 914th AC &W Squadron, Armstrong AFS Canada, part of the Pinetree line in 1960/61.

The liquor we had was the NATO, tax-free liquor. Crown Royal for example was $2.85 a bottle. The most expensive liquor's were $3.15 a bottle. Bartenders were charged with maintaining the bar inventory when on duty and charged .15 cents per shot. They had to account for 28 shots in every 32 ounce bottle. That allowed for up to four shots for spillage and over-pours. The bar manager and Club Officer used the inefficient method of a shot ruler placed on the side of partially filled bottles to reconcile the inventory. So there was some lee-way in the inventory. A1C Lake figured as long as he showed $4.20 in the register he could take a full bottle of bottom shelf booze, which he did.

Master Sergeant (MSgt) Paul LaCasse was the first sergeant with way over twenty years active duty and he did not like anyone who would snub his nose at tradition or orders. Of course with his rank and his position he felt he was exempted from some of the tradition and orders, one of which was having booze in the barracks.

The 914th had rules that alcohol would not be permitted in the barracks rooms. Drinking was to be controlled by allowing it in one place only that being the club. However, MSgt LaCasse was above that law and regularly purchased a bottle of booze for his room from the club manager or the Club Officer. He would never stoop to purchase a bottle from a lowly bartender. There was no secret about this waiver of privilege in fact he would invite some other senior NCO's to his room for a drink when the club was closed or if he wanted company. He was a true believer in RHIP, rank has its privileges. Many traditional old soldiers thought NCO's, especially the first sergeant, had to remain aloof from the lower ranked enlisted personnel to have their respect. That is what many of the old soldiers thought and did. However the USAF was not that traditional because it was so new. The make up of the lower grade enlisted personnel was better educated and an all volunteer force, so rank did not mean that much to them. Age, position and experience were their guide for taking orders. If an older man who showed he knew the job, asked an airman to do something, it was done. Rarely did Air Force senior NCO's have to tell a subordinate to do something, they asked them.

MSgt LaCasse conducted a barracks inspection every Friday morning. Floors had to be clean waxed and buffed, the beds made, shoes and boots clean and polished, uniforms hung neatly and in order in the closet and the top three drawers of a four drawer chest were neatly arranged. He made certain there were no girly pictures or obscene literature hung on the walls. The only sanctuary was the bottom drawer of the chest of drawers. Violation of the rules would mean the violator would have to perform extra duty, or for the serious violations like having liquor in a room, an article fifteen, non judicial punishment and would usually end up with a one stripe demotion.

AlC Ernie Lake, in spite of Armstrong being a remote assignment, had driven his personal vehicle to the site. This variation from tradition earned Ernie the wrath of the first sergeant who made inspections of Ernie's barracks room almost personal, certainly more detailed. However Ernie knew this and acted accordingly. His room was immaculate and in military order.

In addition to being the club bartender, Ernie liked to have a drink on occasions. Not allowed to drink while tending bar posed a problem but not an insurmountable problem for Ernie. Bartenders were required to account for twenty-eight ounces at fifteen cent an ounce for every thirty-two ounce bottle. That allowed four ounces for spillage. When he got off duty at the club he would put $4.20 in the till and take a bottle and hide it in his car. Then he figured why hide it in his car when he has a perfectly good room. But the enemy was amok.

During a routine inspection, MSgt LaCasse entered A1C Lake's room for his inspection. He opened the bottom drawer, the only supposed to be sanctuary, and there he found an opened, partially filled Vodka bottle. He stopped his inspection and then because he was not a clerk, spent the rest of the morning typing and erasing then retyping Article 15 papers charging A1C Lake with violation of Squadron policy. He then finished the chore and smilingly called Ernie to the orderly room. He was going to have the Commander take a stripe. That should teach airman Lake not to thumb his nose at tradition.

Ernie reported to the orderly room and was confronted by MSgt LaCasse. He was told to sign the Article fifteen papers. Ernie refused and requested a Courts Martial which was his right. Sergeant LaCasse was livid. He was all set to work on the preparation of Summary Court Charges when Ernie told him the reason he was refusing to sign. The bottle did not contain vodka. It was water. MSgt LaCasse sniffed then tasted the contents and found it was not vodka. In a fit of rage he ordered Ernie out of the orderly room. He had wasted half a day typing papers that were worthless. It was Ernie's turn to smile.

Another story about this unforgettable character, Ernie Lake, from the author:

When an airman was going on leave he could purchase up to two bottles of booze to take on leave. If an airman was going on a three-day camping outing or on a pass (usually three days without having to work) the airman was entitled to purchase a bottle to take with him.

Two airmen were going on leave and Ernie and two others were going to the Fort William/Port Arthur area to visit a retired civilian from the site. Old Mac was about seventy years old and had worked on the site as a civilian in Power production. He had been a real joy to be around and he became pals with many of the airmen. Ernie agreed to take the two airmen the hundred and fifty miles so they could catch a plane. He gassed up his Ford and loaded it with a few gifts for Mac. The five men started down an old lumberjack truck road the hundred miles to get to Queens Highway 1, a paved road. The first hundred miles had to be driven slowly because it was really only two ruts. Several streams had to be crossed which necessitated stopping and searching downstream for the logs, boards and beams that were the bridge. Finding them they would be brought back to the road to be laid across the stream bed, in essence building a bridge, and then drive across. The next rain would wash the timbers away to be searched for by the next idiot travelling the road.

The five airmen had been driving and building bridges for about three hours when Ernie noticed his gas gauge was dropping faster than it should have. A quick inspection revealed that the gas tank and been punctured and gas was running out. Naturally, the first thing done was to put a finger over the hole while they tried to figure out how to fix the problem. They tried chewing gum stuck in the hole just to have it fall out as soon as the finger holding it in the hole was taken away. Wood pegs were cut and jammed in the hole but it would still leak. The men were out of thoughts except for the one where one was going to have to walk back the fifty miles to Armstrong or forward fifty miles to the junction to get help.

It was proven that day that God takes care of idiots and in some cases Flyboys. Down the road came a car with two Canadian hunters in it. Abiding by the rules of the lonely back roads they stopped and heard the travelers' plight and begging to take them back to the site. One of the Canadian men was about 25 and the other about 50 and they were father and son. They said they were out hunting birds and after hearing and seeing the problem asked. "Do you have any soap with you, you know a bar of soap" The Canadians were assured that the airmen did, but being Air Force, all they had was Dial soap. They said that would do, "Take the bar of soap, eh, and rub it over the hole" Ernie did that and within minutes the drain of gas stopped. Ernie and the others could not have been happier or more thankful.

Ernie profusely thanked the two men and asked if they would like to have a drink. The younger of the two said, "Sure a beer would be nice, LaBatts if you have it."

"No, No, I mean a real drink."

Alcohol was extremely dear in Canada, highly priced and highly taxed and only available in providence run liquor stores.

The younger of the two said "Sure. What do ya have?

"Well we have some rum, scotch, bourbon, whiskey, vodka and rye. What would you like?"

"Ah Bullshit, you ain't got all that," the younger man said. The old man wasn't talking.

Ernie went to the trunk and began bringing out bottle after bottle and set them up on the hood of his Ford which was no longer leaking fuel. Apparently the gasoline and something in the soap hardened and sealed the hole. The hood of the Ford looked like a bar in the middle of nowhere. The two men were amazed, shocked and even envious. They selected bourbon. Ernie asked if they would like to have it straight or mixed. The two men were again amazed when Ernie went to the trunk and produced bottles of mixes every fully stocked bar has. Now you also have to remember, it was summer that day, probably 80 degrees and the gathering was in the middle of a Forest on a two rut road fifty miles in any direction from any civilization. Ernie asked if they would like a glass and ice.

At that point the old man said "Boys, I quit drinkin' some time ago. But today, I think I'll have a little toddy. I ain't never seen anythin' like this, eh."

Ernie fixed drinks all around and for the next forty-five minutes or so the men socialized and had several drinks. Ernie fixed them both with a bourbon and water with ice in a glass to go and they left weaving down the road. The five airmen were also weaving down the road.

After getting the two airmen going on leave to the airport, Ernie and the two friends arrived at Mac's house at about midnight, much later than expected. Banging on the door, Mac's wife wanted to know what the hell was going on and for Mac to tell whoever it was to get lost. When she came to the door and saw three guys carrying two cases of booze, all the mix, an ice chest and glasses, she changed her tune quickly. For the rest of the night the group sat up and drank and talked, laughed and told stories. Mrs. Mac, it was learned, liked her booze too. She fixed breakfast and then she fixed lunch and then dinner. She was having the time of her life.

It had taken Ernie and the others several months of buying bottles of booze, twenty-eight shots at a time for .15 cents, a total of $4.20 a bottle to accumulate as much liqour as they did and not attract attention. It had all been hidden in Ernie's car with the appropriated mix and glasses. Mac was a buddy and they wanted to take care of him.

When the group left Sunday morning to return to Armstrong, left was at least fifteen unopened bottles of booze and a couple cases of coke and seven-up along with a dozen or so used dirty glasses and two Canadians with hangovers they would remember for some time.

Ever since then whenever I travel I carry a flask of booze in case I run into a Canadian.

KIDS AND THEIR CARS or BOYS AND THEIR TOYS

Men take pride in their cars. Military men during this time took great pride in their cars mostly because they drove what they could afford which often would be motorized junk. But it would be their motorized junk not GI (Government Issue).

Mike Atwood:

We had a 2nd Lieutenant who was a good guy and would party with the enlisted men, especially his crew members. He really did not deserve this but a couple guys happened upon his 1954 Chevy parked on the street in Mill Valley, the town at the base of Mount Tam

151

California. One of them had a ladies Bra from an earlier happening that evening. They tied it to the antenna of the Lieutenant's car. He did not notice it when he got in his car and drove back to the site. The Air Policeman said nothing to him as he passed through the gate upon entering the base. However the next morning the Commanding Officer, who was the best CO I had on any of my assignments beckoned the young lieutenant to his office and told him it was not what a gentleman should be doing, driving around with ladies undergarments streaming from his radio antenna. The Lieutenant was puzzled as he still was not aware as to what the CO was talking about until he took a good look at his car.

Bob Wildrick:

My 1st car in the Air Force was a 73 Subaru. I got "blottoed" in the NCO club at St Albans AFS, Vermont one night and offended a few of my fellow fly boys. Snow had fallen and there were a few inches of snow on the ground. I went out to my car a while later, NO CAR! Old drunk me went back into the club and demanded, "Where's my car? I left it right outside." Needless to say I received no answer. Later I found it a 1/2 mile down the hill, sitting on home base on the softball field. Four or more of them had picked it up, slid it down the hill on the snow and then slid it over the field to home plate. I GOT the message!

Cliff d'Autremont, A1C (E4) USAF 1962-66

I was at Grand Forks AFB, North Dakota 1964-66 and when I first arrived I bought a 1960 or so Ford Taunus which was really a German Ford. It was about the size of a Toyota, much smaller than the big American cars of the time and I loved it. Only problem was getting parts. They had to be ordered from Germany through the local Ford dealer. Typical delivery time was four to six months. I was able to get a muffler fitted without waiting for the long delivery time but in general, parts were not interchangeable. It had a four blade radiator fan. One day one of the blades broke and I ordered another one. It was going to take four months. I thought I could still drive with three blades since it was winter. My thinking was wrong. A few weeks later another blade snapped off due to the fan being unbalanced. But the second blade that snapped off was opposite the first one which had broken. Now I had a balanced two bladed fan and continued to drive. Weeks later another blade snapped off, and then the last one went. But I was OK as long as I didn't stop moving too long.

I was riding out in the country one day and there was a snow drift in a low spot on the road. I thought, *I'll just pick up speed and plow through the drift and come out the other side.* Again my thinking was wrong. I got stuck. I could rock back and forth and make a little headway but the engine would overheat since there were no fan blades and I wasn't moving much. I'd have to wait a half hour for the engine to cool and then I could rock back and forth for a short while and then wait some more and so on. It took over 4 hours to get out of there.

I traded the car to another airman for his beat up 1960 Chevrolet. He probably got the better of the deal because the replacement fan on order would arrive but it would be after my enlistment was up.

Lowell Woodworth, On getting by without traffic tickets:

I was stationed at Empire AFS outside Traverse City Michigan and my girlfriend, soon to be wife lived in St

Albans, West Virginia (Oops, that s West by God, Virginia). To get there I had to go through Ashland Ohio. The town cop had a reputation for strict law enforcement so I had to be extra careful. The Ohio State Patrol also had a thing about loud mufflers and my car had smitty mufflers. The only way I found to quiet the car down was to pack the muffler and tailpipe with steel wool. We had plenty for cleaning purposes in the barracks and with a broom handle I would stuff the muffler and take off. I could usually get to Portsmouth Ohio before the steel wool burnt out and the "smitties" would come to life. From that point on it would be a slow run to the border. I guess I was lucky I didn't burn out the valves, because I had to repack the system to get back to Traverse City. Ah the things we do for love.

Raymond Raflik added this story:

It was 1957 when I was transferred from Hokkaido Japan to Charleston Air Force Station (CAFS), Maine, and located somewhere near Bangor. I was Airman 2nd class. After landing with a commercial airline at the Bangor airport, I was shocked when everyone I asked did not know where Charleston was. Most thought I was supposed to report to Charleston South Carolina. My orders said Maine, so I slung my duffle bag over my shoulder and headed down the nearest Bangor Street. After several blocks, I approached a feed mill where a farmer was loading sacks of feed in the back of his pickup truck. I asked him if he knew where Charleston Air Force Station was located. In his Maine accent, he replied, it's about 30 miles up the road and he was going right by there. If I wanted a ride he said to throw my bag in the back of the truck and get in. The farmer dropped me off and I signed in at the orderly room. I was processed in and then went to the club to pass the time.

There would be many hours spent passing time in the future.

I had been on the base about two months when I took a 15 day furlough to attend my sister's wedding in Wisconsin. I hitchhiked back to Wisconsin to save money and bought a three year old, 1954 Chevrolet. I drove the 1540 mile trip to Maine and arrived in Dover-Foxcroft about ten miles from CAFS. I wasn't sure which road to take to the base so I stopped and asked the elderly police officer and got the instructions. After I drove off I saw him jump into his 1940 Buick police car which did not have red lights or siren and come in the same direction as I was. I stopped at a stop sign and he passed me, pulling in front and stopping. He told me I was driving an unregistered vehicle because my car did not have license plates. I informed him that I had just purchased the car and the dealer in Wisconsin had sent for the registration to Madison Wisconsin and that the stub from the money order was taped to the windshield. The date on that stub was handwritten and was 7-15-1957. He thought that the seven was a one and instructed me to follow him to the courthouse. I explained to the Magistrate that I was stopped twice on my trip and after explaining my purchase was told to continue my journey. I found out later that the Magistrate didn't like servicemen and he ordered me to pay a $25 fine and $7.50 court cost. The final blow was him ordering me not to drive the unregistered vehicle. It had to be towed. My plates arrived the next day and I reported the incident to the Provost Marshal at CAFS. He started making contacts to clear my name from two newspapers in which my fine was reported and to get my fine money back.

Stories were floating around that when I got through suing Piscataquis County for false arrest, they will probably re-name the county, Raflik's County. Of course, for the sake of the USAF, it never happened.

In August 1957, a new tile floor was installed in the hallway on the 2nd floor of our barracks. Whoever installed the floor left a gallon can of turpentine near the staircase. Someone came in from town and kicked that can spraying turpentine all the way down the hall of that new floor which then bubbled-up and had to be replaced. Because no one admitted to doing this deed, the whole barracks was restricted to the base for three weeks. My room was on the first floor so I certainly didn't do it but my pleas went unheard. Clearly this was another miscarriage of justice, so in quiet retaliation when I was off duty I would get in the trunk of my car and have a friend from another barracks drive me off base. Having that car caused me a lot of trouble. I didn't have a driver's license due to another incident where I had been drinking at a party and I felt I was too drunk to drive so I pulled to the side of the road and went to sleep. A highway patrolman came by and gave me a drunk driving ticket. The judge found me guilty and my license was suspended. That did not stop me. I continued to drive my car after someone else drove it off base.

I considered myself a good driver even in snow and ice. Others were not. One Friday evening in October 1957 several of us were on the way to a dance. With us in the car were eight 48 ounce bottles of beer. It started snowing very hard with large flakes. Visibility was not good. I started to climb one of the steep hills when I noticed something up ahead in the road coming towards us. The object turned out to be a pick up truck being driven by an elderly lady. She was backing down because she could not make it to the top, had given up and decided to back down. The trucks had very small taillights and were covered by a homemade wood bumper. She hit my car and smashed the grill and left light assembly. After checking if she was hurt I asked her if she knew anyone in that area in order to call the police. She said she did and she went to make the call. That gave us time to bury the eight bottles of beer in the ditch.

One of my passengers, a kind of a loud mouth on the base, said he would say he was driving because of my not having a license. That is how the report was filled out. A couple of days later, I was called into the Provost Marshal's office and asked about the accident and why I hadn't reported it to the base? I said because it was a minor accident with less than $100.00 damage I didn't think I had to. He then asked me if I was driving. I had to stick with the report as filed with police so I said the other guy was driving. The Provost Marshal said this other guy just left my office and said I was driving. I now was in deep trouble. The Provost Marshal threatened to write to the state of Wisconsin and have my driving license revoked for 5 years. After explaining each of the incidences that I was involved in, he had me write a statement. Then he told me he was impounding my car and I had thirty days to get rid of it. My brother and my dad took a bus to Maine and drove the car back to Wisconsin.

I had grown up on a farm and was an avid hunter. The deer season in Maine was the entire month of October so I purchased a deer license. A friend wanted to go deer hunting with me but he had never been in the woods much less hunted anything before. We had no idea where to hunt so we took a highway north of Milo, Maine into some very wilderness country. This area was heavily wooded and there was a small stream down in the valley with a deer trail along both sides of the stream. I posted my friend, crossed the stream and found a spot for me to sit. After several hours, as evening approached a small buck was making his way along the deer trail. I shot the buck in the neck and noticed several other deer run in the direction of my friend. I heard several shots fired and assumed he also shot a deer. I quickly gutted my buck and went to help my friend. He said a large buck ran by him but he missed it. We agreed to get my deer out before it got too dark to see but due to all the deadfalls it became impossible. When night fell, we decided to leave the buck. We covered it up with brush so animals wouldn't eat it and to come back the next day to retrieve it. There was some discussion about which direction my car was. Neither of us brought a compass. I spotted a group of stars shaped like a diamond in the sky that pointed in the direction I felt was right. As we walked, several times my friend wanted to go a different direction, but I insisted we follow that group of stars. After a time we came out about 100 yards from his car. Back at the base my friend told everyone I was a Daniel Boone, coming out of the Maine wilderness by following the stars. It became my nickname for the rest of my hitch.

Bernie Morris. Served on active duty during 1954-1958:

When at K-18 Kangnung Korea in 1957 we had a lot of spare time on our hands. Two young troops, just out of mechanic school were assigned to motor pool and decided to fill their idle evening hours doing something constructive. We had several damaged vehicles on the site in a junk yard. They cannibalized the wrecked jeeps and with the good parts, ended up creating the most beautiful Air Force jeep you have ever seen. They spent several evenings painting and repainting this vehicle rubbing it down between coats. The put air horns salvaged from trucks in the junkyard on the front fenders. Some of the local's who worked on base even took some parts to town and had them chromed. The upholstery was done by the base tailor. Because this was done using all government owned parts which were in effect stolen, it did not have any registration numbers painted on the hood and bumpers. The commander restricted it to on base use only. The commander felt this vehicle was such a showpiece and he confiscated it for his own use. Needless to say these two young Airmen were devastated upon losing their much admired and accomplished vehicle. The Commander drove the Jeep to Osan K-55 on all day trip which was his big mistake. He left it parked and Wing Commander, a General, at Osan saw it. He decided this should be his vehicle and relieved our commander of the beauty. I was at Osan soon after that, and as I was walking along the road, lo and behold here came the General's jeep in my direction. General flags flying from the fenders which were still adorned with those shiny chrome horns.

In honor of my buddies who put this beauty together I stopped as the General passed and saluted. I was not saluting the General; I gave the Jeep one big high ball salute. The General never knew.

Dean Gee. On ingenuity:

It was early winter at the radar site in northern California. Two of us decided we would go down the hill into town and get a hamburger. On the way down it started to snow just a little. We ate and went back to my old 51 Studebaker with the broken windshield wipers. Surprised, we found a few inches of snow piled on the old car and more coming down. We knew the cops would cite us if we didn't have a clean windshield so I went back to the restaurant and sweet talked the waitress into some string. Tying the string to the wiper arms and bringing it into the car through the side windows, we could pull on the string and make the wipers move back and forth just like they should work. Now we could get back to the base at least so off we went. My heart went to my throat when I saw the red lights flashing behind me. I knew I was going to get a ticket, the car would be impounded, and we would have to walk up the hill because no one in their right minds would be out in such a snowstorm, except us two fools. I pulled over and dug out my license and registration and insurance while he ran my plates. I tried to make certain my security Police badge I had put in my wallet would show and he would give me some professional courtesy, otherwise I could kiss my next paycheck goodbye. The Highway patrolman approached and fortunately he did not ask, "Do you know why I stopped you?" I would have said, "because of my windshield wipers?" Instead, he said, "I stopped because you have a tail light out." Inwardly, I breathed a sigh of relief, signed the repair order ticket and bid the officer to "be safe." Fortunately he did not see the badge or the string. I found out later he was the kind of cop who likes to make an example of other people to show his total objectivity.

Another time another buddy had an old wreck of a car not worth a wrecker to come and tow it to a junkyard. Everything was wrong except for the motor and it just barely ran. The brake's master cylinder worked but the hoses had so many leaks if you had a full cylinder and put the brakes on the car might stop the first time. After that there would be no fluid in the system. New brake fluid was expensive and my buddy did not want to put out

160

anymore money in fact if he drove the car to the junkyard he would get $50. His solution was water. He filled a jug and the master cylinder and off he went. He got there in one piece after stopping several times to put water in the brake system.

Isaac Thompson provided this:

Some of the guys had some pretty hot cars. They had time to soup them up for speed and really liked to drag race. They would pull up to a red light and when it turned green the race was on.

While at the 683 AC&W squadron at Sweetwater, TX in 1966, an airman had a 64 Chevrolet Impala Super Sport with a 409 engine and oversized tires. Another airman had a 1963 Chevrolet with a factory special 327 engine. These two cars were the hottest in four counties. Every cop knew the two airmen and also knew the two cars. But the cops had never caught us drag racing anyone. Then one night I was stopped at a light. The street was dark and this car eased up along side me and began revving his engine. The signal for a challenge and I was ready. The light turned green and I was off. The challenger was left in my dust until he turned on his red light and siren. My heart was in my throat and I was visualizing what the inside of the jail looked like. I pulled over and he pulled up red light blazing. He came up to the car and shouted "Henderson?" I yelled back "No, Thompson." He came up to the side of the car laughing and said "I heard you guys would race anything, now I know it's true."

He didn't give me a ticket just asked me not to do any drag racing in town. I went back to the site and vowed that next time I was challenged that I would look at the entire car before I accepted.

ON SPORTS

Sporting activities were a favorite pastime on the sites. Some participated and some only watched but they all enjoyed the diversion.

William Seiter:

We had no entertainment at Saratoga, New York in 1951 to 1954. We didn't need one. We had lots to do around the lake at the local bars, Night Cap, 656 club and Jack and Irene's. We did have a fast pitch soft ball team and joined a league sponsored by the race track. The race track also held boxing matches. The winner was awarded $20 and the loser got $10. Anyone could fight. There were a lot of dives taken for beer money around the end of the month.

William "Bill" Dougherty described Mount Hebo Oregon as follows:

Mt. Hebo was on top of a mountain overlooking the Pacific Ocean. Six miles up the access road were 27 single family homes for use by key enlisted personnel and the officers assigned and their families. Driving the additional two miles would get the traveler to the top of Mount Hebo and the radar site. Two four wheel drive school buses were available for the civilian dependants from the housing area so they could get to school. GI's were allowed to travel by these busses as well if there was space available.

During the winter when there was snow on the ground, and that was about nine months out of the year. Many of the men who lived in the housing area could slide down the access road on cardboard or plastic sheets, getting home without taking a step.

Snowshoeing was one of the favorite winter sports the troops enjoyed along with skiing, skating, sledding and tobogganing. Some of the lakes would freeze and became ice and hockey rinks.

When not on duty, there were many activities to occupy their time. A small library, theater, gym, snack bar, bowling alley, and hobby shop would help pass the time.

Wayne Fitzgerald:

I don't know about the other Pnetree sites but we had a curling rink at Beausejour which we built ourselves. Several good teams visited the Bonspiels around the area and always had a great deal of fun. Edmund A Manuseto and I curled together calling ourselves the Hard Rockers. We used no finesse, no fancy sweeping and curling, just slide the stone with great velocity and knock the other guy's rock out of the bull's-eye. We did not make lasting friends that way but we beat lots of the good precision curlers and that was good fun for us at least. Curling and hockey were really popular in Canada and I'm certain other sites had curling rinks too. We also built a hockey rink at Chandler and had lots of fun playing hockey until my buddy Bobby Nama about killed the goalie. Our Commander Major David Jones was the goalie and Nama did a slap shot which caused the puck to go right through the peep hole in the mask. Major Jones went around looking like the Lone Ranger for several weeks. I have never before or since seen such a pair of shiners.

Lawrence Kessler. On baseball:

In 1957-60 we had a baseball team at the 745th at Duncanville, Texas. I was the catcher and had a good wing but couldn't hit the ball, still we were pretty good. We got better when a Technical Sergeant named Ed Tatyrek was assigned. Before entering the USAF he had been a pitcher for the Houston Buffs a triple A team for the Cardinals.

While we did not have enough players to go on the road for intramural competition we did play some local high school teams and with Ed's pitching we "Moidered da bums." He had the best curve ball I have ever seen and to this day, I swear, it moved three feet. It didn't help my batting but I sure enjoyed catching for him.

Paul Coutu:

We arrived in October, 1953 at Hopedale, Labrador. The base was brand new and bare. We brought with us all of the furniture and food. There was no radio station, no bowling alleys or any other type of recreation other than a pool table and table tennis table which we also brought. You can imagine what happened to the pool table and tennis table. With play going on 24/7 the pool table had ruts in it rather soon. Needless to say the ping pong balls disappeared rather quickly. We went weeks with no communication with anyone. Mail was something that came only when weather allowed. Some of the conditions I have read about here would have seemed like a cakewalk to me.

Ray Vena:

Here is a funny story about a site recreation director, an airman second class, who wanted to play on our site baseball team.

He had never played ball before but he ordered a brand new glove and asked us how to break it in. We told him to put a ball in it, wrap a large rubber band around it & soak it in a pail of water overnight. Well, the next day when he took it out it was like a cement block. He was devastated, but we got him a new glove and made him a member of the team. Not all personnel assigned to that recreation director position knew a whole hell of a lot about sports.

Joe Sargent:

Ah yes, curling at Finland AFS, Minnesota. I was a new RCA representative back in 1961 and it was my first winter in Minnesota. It was also the first time I saw a curling rink as well. Newly arrived Technical Sergeant Bill McVitty and I went down to see what it was all about. Both of us almost fell over at this foolish game and we questioned what it was all about. They threw a rock on the ice then used a broom to sweep in front of it. Why not just throw it harder and faster and throw sand on the ice? It looked so foolish to us and of course we had to start to tease and taunt the players. After enough ribbing, the players invited us to give it a try if we thought it was so easy and foolish. Since I knew bowling and a bowling ball weighed 16 pounds, that little rock couldn't weigh more than that. So I accepted the offer and hefted the 42 lb stupid rock with its handle and figured okay it is a little heavier heftier. I tried a few bowing approaches minus the stupid rock just to get the feel of the ice. Then I gave it a live approach with the stupid rock with just enough heft to get it as far down the ice as I could. Needless to say, it zoomed down the ice, hit the back, propelled itself airborne and imbedded itself into the sheetrock wall. Nice goin' A$$&0*#.

McVitty and I eventually got hooked on the Stupid game, got into a league and got pretty good with a lot of help from Canadians and Minnesotans'. McVitty got to where he had an uncanny ability to read the ice as the games progressed and the ice characteristics changed. He was great as a team captain and he would have us putting stones around blockades like magic. We even won a few trophies at some Bonspiels. We also drank a lot of Crown Royal to keep warm (as if we cared about keeping warm). Whenever there was a Bonspiel at any of the Canadian sites or in towns close to a site many of our guys would go up to cheer us on or to play. We always remembered to bring back the maximum allowable liquor amount. If I recall right, Crown Royal cost me $1.19 a fifth from the site NCO club. Ah yes, Almost 50 years later, I think fondly of that stupid game, of those stupid rocks with handles, the hours of fun I had and the many friends I made.

"Doc" Mike. Stationed at Calumet, Michigan:

Winters in the Calumet area of the Upper Peninsula of Michigan were long and hard. We had snow banks that were higher than the tops of the telephone poles on Highway US 41. The norm was 300 inches of snow. What do you do in the winter, well we learned you play snow and ice games. One game we learned was hockey. A group of us airmen formed a team and practice every chance we could. We were skilled enough to compete. We challenged a local team figuring we could have some fun and learn more about the game. We did. We got our butts beat by a team of grade schoolers.

Bob Moore. Wrote:

I attended Squadron Officer School back in 1968. We had to make several five minute speeches. One of our classmates made a powerful speech about the importance of an officer always being in proper uniform. He emphasized the importance of "attention to detail." After the speech, the rest of us would critique the performance. The first question asked of him was, "is the fact that you are not wearing a belt meant to be an "attention getter?" No one had noticed until he made his presentation.

When I was a young Airman second class at Ramstein, Germany in 1958, I worked with a Staff Sergeant who told us something once that really made us laugh. He told this on himself so it must be true. This man had been prior service with the Army as a Corporal when he came into the Air Force as an Airman First Class. He was at Lackland and had been issued all his new uniform clothing. He went back to the barracks to get ready for the next day. He told us of how he had sat up most of the night sewing on stripes and pressing uniforms. He got up the next day, dressed in his fresh khaki uniform and strutted to the mess hall. Upon entering he was immediately jumped on by a Master Sergeant. He had spent all that labor getting his uniforms ready but had sewn all the stripes on, upside down. He must have thought he was still in the Army where the stripes are sewn on with the point up but in the USAF, rank stripes are sewn are sewn to look like wings, inverted.

We were young and not so smart when stationed at Patrick AFB Florida and assigned to the 645th Radar Squadron. Some of us had imbibed a bit too much at one of the local dives and decided we needed a swim, a little dip in the Atlantic just across the street to complete the night. Of course we had no swimming suits so it was to be a skinny dip.

I was the least drunk of the lot and decided I'd just watch and make sure nobody drowned and signal if anyone showed up. After a few minutes watching to make sure they were sober enough to swim, I told them I was leaving and walked across highway onto the base.

There is a strange and yet wonderful thing about the ocean. Tides come and go and things you leave near the tide-line, such as clothes, just get washed away. So there they were, drunk as could be without any clothes or military ID and it is 5:00 AM and the sun's coming up.

Somehow, they convinced the on-duty Security Policeman at the gate that they were airmen and to please let them streak onto the base, each wearing half of a cardboard box the SP loaned them. Where's a camera when you need one? Oh, and they blamed me for their clothes being missing.

Ernie Bickford:

While stationed at Otis AFB, Massachusetts '55-'57 I was to receive the Airman of the Month award sometime during the summer of 1956. Well..., Falmouth was a wonderful place to be stationed during the summer and one thing led to another. The nite before the award I was in town with my cousin and after consuming a few we went to the local hangout for something to eat.

Somewhere along the line my chair got pulled out from behind me and I did what I had to do, I slugged the guy. Cops came and I spent the nite in the hoosgow, getting out in time to report for the award. The trouble was the First Shirt already knew about the episode and that was the end of the award. My parents and friends from Massachusetts had travelled the two hours to attend the ceremony and sat there wondering what the heck happened. Oh well, what could I say.

The guy sued me for $600 that I had to borrow. The day I went to pay his lawyer I had pulled $5 from the wad to have some beer later, then claimed I must have mis-counted and that was all I had. He took the $595 bucks and I never heard another thing. I went and had a few beers to soften the blow of giving that much money away.

Bob Moore:

I was stationed at Tin City AFS, Alaska in 1961 through 1962. There was little or no recreation at this time, unless you want to consider movies where Joan Crawford was a teenager. We were shown that two or three times a week until weather permitted the landing of an airplane with a new film. There was no bowling alleys, no television, Armed Forces Network radio was sent to us via microwave, which at best was received over a lot of static. Our days consisted of getting up, going to work, getting off, going to the bar, drinking ourselves silly, going to bed and getting up the next day and doing it all over again. I can remember card games which never ended, just changed players as one got tired or had to go on shift. One spent many hours just walking the hallways and trying to overcome the boredom whlle listening to the wind howling outside and what seemed the never ending

169

night. Maybe I am a bit bitter at what those who came after had which we did not, but, we endured and made it through. I feel those guys who served on remote AC & W sites in the early years paved the way for those who came later. I realize that griping is a way of life in the military, but consider yourselves lucky if you had a TV to watch, even if the shows were late, just think of those who had none.

Bob Fogarty, agreed saying:

Right on, I was on the hilltop at Tin City and there was even less to do up there. We sat around in the kitchen drank gallons of coffee, played cards and listened to the same records over and over when off duty. Sometimes we played bumper pool but we got so good at that, no one wanted to play because it wasn't a challenge any more.

Leo Mulligan says:

Gee and I thought I had it rough on Kwajalein Island in the Pacific, to support the Atomic bomb tests at Eniwetok. We worked six hours on and 12 hours off and after every four shifts we received a 36 hour rest. While it was not an A C & W site, we also had old Tim Holt and Hopalong Cassidy movies which we watched at an outdoor theatre during monsoon season with our raincoats, shower clogs and pith helmets. There was no TV or phones. We did have a day room with three pool tables and a broken ping pong table which we shared with the Navy and Marines, 1300 men, too crowded to enjoy. There was no drinking until you were 21 years old. The highlight of a week would be to see a stewardess get off an aircraft that had problems. I was then sent to Saratoga Springs AFS, a site very near to one of the best resort places in N.Y. The young ladies were plentiful, as the young men had all left for better paying jobs, or joined the Military. We had a speed boat for water skiing, a club for all ranks on station and the movies downtown discounted if in uniform. I thought all radar sites were heaven on earth.

Bill Parsley:

It seems that a few of the remote sites tried to bring up their own radio stations. When we moved the 920th from Argentia to Resolution Island, Canada, we did the same. Jim Tanaguchi, a civilian technical representative and someone else built a small transmitter to use. We received turn tables and large records from Armed Forces Radio Services and set up a station. Our little Base Exchange branch sold small Arvin radios so most of the guys without a radio bought one. We were also piped into the mess hall so when we were on the air you could still hear us. When we did go on the air, we knew the frequency might have drifted a bit so when a DJ signed on, they would sign on with "This is station WROCK being brought to you with an unknown power output and an uncontrollable frequency." We figured if we were putting out 1/4 watt that would have been a lot. There were a few of us taking turns as DJs so everyone would play their kind of music along with giving any news we got. Some of the guys made up their own commercials and announced them. One of the best was "The preceding program was brought to you by the Pikes Peak Brassiere company whose motto is, together we stand, divided we fall." We had a blast.

When I left Resolution in April 1955, we had to fly up to Frobisher Bay over-night before heading down to Goose Bay the next day. I ran into a friend who was stationed there and when he heard I was from Resolution he asked me, "Are you that station WROCK". When I said yes, he responded that it was a great station and a lot of the guys there would rather listen to us than stateside, they didn't know what to expect. It almost knocked me over to hear that they picked us up after all; they were a couple hundred miles north of us.

Later I was stationed in Roswell New Mexico and met someone who just came back from Resolution. I asked him about the station. He said that the Canadian Government found out about it but allowed it to stay on the air after receiving good reports from their citizens.

J. Lorenz was in the USAF in 1950 to 1954:

While stationed at the 755th Williams Bay, Wisconsin several of us took a drive up to Milwaukee and visited the Schlitz Brewery. Their slogan was "Schlitz, The beer that made Milwaukee famous." It was very interesting but of course not being 21 yet, I couldn't sample the wares (yeah, right). Glad to hear they are still in operation.

ON THE LACK OF WOMEN AND DEALING WITH FEMALE PERSONNEL

Former male and female Airmen share their experiences with the opposite sex while at the sites.

Ray Bergen, about using the phonetic alphabet and supervising females:

When at the 635th at McChord AFB in 1958-59 we not only had WAFs but had several Civil Service women also working in Operations. I was a Floor Supervisor and had at least two WAFs and three CS women working for me.

172

There were also some male CS employees as well. They were restricted to working the scope, plotter and teller positions while the WAFs were able to train in identification and target intercept spots. When receiving information from another site the information was always acknowledged by giving the initials of the recipient using the phonetic alphabet.

One of the WAFs was Charlotte Starnes who would always vary from the proper and gave her initials as "Candy Sugar" when talking to other sites much to the chagrin of the Directors. One night the Senior Director (SD) asked where she was and I told him that I had giver her the night off because she had called me and said she wasn't feeling well as "it was that time of the month." The SD said didn't you give her a night off only a couple of weeks ago for that same reason? When I said, "yes" he said, "Ray, I want to talk to you." Heck I was only 25 and single and didn't know about things like that. After my biology lesson from the SD, I called the WAF barracks and told her to get to work in 15-minutes or else. Needless to say it was a long time until Charlotte got another night off.

Another female named Lynn Paulson was very happy when the phonetic alphabet was changed. She could then acknowledge conversations with Lima Papa rather than Love Peter.

Ed Simpson, MSgt USAF Retired:

Those were the days when men were men and girls were girls, and the booze was running like water over the Niagara. These are memories that only us OLD Airmen can relate to. Yes, the 50's and 60's were the best years of the Air Force.

An unidentified ex-airman responded to Ed Simpson:

Ed, you should have served in Saglek, Labrador where men were all there were and girls were 600 miles away available only by the weekly plane if you could weasel your way onto it. The booze, admittedly only 10 cents on special nights, didn't quite make up for the tears of frustration related to - oh yes - the girl issue. Ah the 50's and 60's - I remember them well!

Another unnamed airman provided this story:

I had a bad day at Palermo AFS, New York, and needed to get some quiet time and a beer. I left the site and went to an out of the way bar with cheap beer and a good jukebox. I felt a little awkward and out of place as most of the other customers were officers. These guys were Navy and Air Force pilots and always looked good in their uniforms with their wings, short hair and California tans. I'm a little more than nineteen, not socially adept and a little shy of the required twenty-one required for a legal drink.

I was working on my second beer when this "beauty" walked in to the bar. All eyes were on her and her mother. Mother looked a little tipsy and not, shall we say, a picture out of Vogue.

I couldn't help myself but to stare, I had never seen such a beautiful woman. I thought to myself, what could I do? The Officers had the looks, the cash, and they could dance better than I could.

After my third beer, my courage prevailed. I'm gonna ask her to dance and then my courage failed. God, she was beautiful and beautiful in all the right places.

The officers were hitting on the "beauty" for hours. What am I going to do? I had been taught by the Air Force to think strategically. So I got up, walked to her table and asked her Mother to dance. Mother looked up and said "you want to dance with me?" I said why, yes I do.

We finished the dance and I thanked her and returned to my beer at the bar. I finished it and was getting ready to go when the "beauty" came over to me and thanked me for being so kind to her mother. She asked me if I could help her get her Mother home.

What do you think I did? Did AF training payoff?

Fred Boutin, Airman First Class:

I served at one of the remotest isolated sites on the Pine Tree Line, Saglek AFS Labrador and I am one of the ones the Air Force spent a fortune on training and lost their investment. There was no way I was going back there or any other site like it. The scuttlebutt was once you had survived the isolation without cracking up there was a list that you went on. It was from that list that they selected airmen for new remote assignments. Likely not true but more than one story has borne it out.

We were rough, sometimes rowdy, and drank too much sometimes and we once had an unofficial contest to see how long any conversation could go without having the "F" word creep into it. The average was about a minute. Not the kind of place to have women one might think, well you would be wrong. I for one think it would have been a much better place and more civilized with women there as equals. As for sex, well it happens no matter how many try to ignore it. The old school Air Force chauvinists made a big mistake in not utilizing equally both sexes. Hell I just might have stayed the twenty.

Nancy Few, a WAF was assigned to the sites:

The radar operations and maintenance fields were opened up to more enlisted women in the mid 70's. When I went in 1971, it was not an option, and I was stuck in personnel as a clerk. In April, 1974, I cross-trained into one of the electronic fields and there were a few other women in my class. I was the second WAF at Almaden AFS, California the other being a cook. They didn't have quarters for women so we had to live down the hill. I was told to settle in, that I was frozen there for three years, so I put in for lateral training into SAGE. They had a class date for me and ADC was telling me to pack my bags. They were just waiting for my Date Estimated Return form Overseas (DEROS) waiver to come down. Since I had just taken an early re-enlistment for 6 years, and hadn't been overseas, the DEROS waiver turned into orders for Rockville, Iceland. I was one of the first of several WAF to be assigned to Rockville AFS, and I presume, remotes in general. We had to live on main base at Keflavik, and since I was a sergeant with over four years in service, I was able to take my car.

I loved going to Reykjavik on my days off, or to Gullfoss, Geysir, or Thingvellir, and to the Icelandic Symphony. I would occasionally take some of the newer airmen with me, just so they could get off the base for a while. Many of the airmen had a bad attitude toward the Icelanders, calling them "Mo-Jacks or Mo-Jenny's" which were derogatory terms as if it were the citizen's fault the airmen had been assigned to Iceland. I loved the Icelanders. Of course, being half Danish and a quarter Swedish didn't hurt. I guess I looked like an Icelander and the locals would speak to me in Icelandic, until I told them I was American.

I got orders to Mt Hebo AFS, Oregon, and wasn't sure that I wanted to go from a remote tour in Iceland to an isolated tour in the United States, so I requested an extension in Iceland, which was approved. I spent 18 wonderful months at Rockville AFS, in Iceland and then assigned to McChord AFB Washington for my last four and a half years. I met a former enlisted Air Force Academy cadet, and soon we were engaged. He was going to be a navigator and counter-counter measures officer, and we thought it would be fun for one of us to be on each end - me on the ground, him in the air. I received a Dear Jane letter and that was that. My next orders were to a Radar Squadron in the northern part of Germany. The location was in the swamps near the Netherlands, and the squadron was located on an Army Post. It would be a three year unaccompanied assignment. By that time I was a single mom, with a three year old daughter. With great regret I refused the assignment and got out with a total of nine years, four months and 24 days in my beloved Air Force. I went back to school at Arizona State University and received a BS in Psychology while spending six years as a medic in the AF Reserves. I guess I could have re-enlisted after I finished my degree but by then I was just over 35 years old and over the age limit for commissioning.

Tom Terrell,CMSGT Retired:

Many WAF personnel were in radar operations later on, however in my early days in the 50's and 60's no WAF were sent to any remote or isolated sites. That meant that only males were being sent remote and the WAF personnel were exempt from that duty. Maybe that is why I wound up with four remote and, or/isolated tours in my 27 years in the USAF.

Donnie L Shahan:

I was stationed at Oklahoma City AFS, Oklahoma and had met some of the local girls. One family with three girls lived nearby and I and some of the other young airmen used to visit them. We would play cards, checkers, whatever, and sit around in a small group and just gab. One of their girl friend, all of fourteen, would be there sometimes. She became a pest right quick because I couldn't get away from her. Wherever I'd be she would be right there. She seemed shy and didn't talk much but she always would stand or sit near me, no matter where I moved. She always wore cowboy boots, poodle skirts with a number of those stiff petticoats, and had a pony tail that almost hung down to the top of the boots.

I noticed that she hadn't been around for a time and was told she had been gone for two years after her and three other teens had been apprehended in a stolen car. She was not prosecuted because she didn't know it was stolen. She went to live with her former preacher's family in Arkansas because they had a daughter her age who was her friend.

I drove one of my friends to his uncle's house one day. The Uncle asked his nephew if he had seen that little Helen girl since she'd come back from Arkansas. He remarked that she was a pretty little doll. I made a mental note to see for myself if she was pretty. I had never thought she was. I went to her house to see her. The uncle had been right. She was gorgeous!! Very pretty face, had a hair cut which was nicely permed, 36"x24"x37". Woo, Woo!

We married 20 Oct '58, but like so many other marriages, it didn't last.

WILDLIFE

It was quite common for radar sites on hill tops to have attracted lots of different wildlife, animals that is, from rats and rattlesnakes to moose and bear. Some share their experience.

Jim Guy:

I only spent 1973-74 at the Top Camp of Indian Mountain, Alaska and when I look back on my experience it wasn't so bad.

Yes, it was cold and snowy and stayed dusk for part of the year, but the summer was actually pretty nice as I remember. The mosquitoes were terrible especially at bottom camp. During the winter we usually had to ration the water since it was pumped up to us and stored in a big tank and toward the end of winter we were allowed one shower a week and the flushing of urinals was a no-no and was assigned to an individual on a weekly basis or so.

We had a lot of camaraderie and the NCO Club was the only place in town. The only way in or out was by plane and that is how we received our food, mail and beer. My winter there was a bad one and supplies usually were not on time so we ate a lot of cold cuts some of which were green. But when we had our supplies, we ate like Kings. Steaks were on the menu most of the time. I'm sure there may have been worse sites, but Indian Mountain had to be in the top five. On the other hand I was in my 20's and really didn't mind the inconveniences and where else, after a hard night of drinking, could you walk to the side of the Mountain and feed the Black Bears cookies for dessert, while they were dining at the site dump,

Pete Holland:

Late one night in the search tower, I was talking with one of the radar maintenance shift guys in the tower office, and I noticed a paw reach out from a hole in the wall to snatch up his sandwich sitting on the desk. It turned out to be a ring-tailed cat.

Rick Chinn:

One day someone sent one of the newbies (we called him, "The Boot") to the storage room around the back of the Ground Air Transmitter Receiver (GATR) building. He felt something strike his combat boot, and looked down to see a rattlesnake. Needless to say, he levitated himself out of there. Since we had classified materials on site, there was also a loaded pistol, and the rattler found itself the object of some impromptu target practice while the boot danced, hot footing it up and down and shooting at the snake that probably was just as scared and left the area amongst the bullet holes.

Unnamed contributor stationed at 814 AC&W Squadron, Killeen AF Base, Killeen Texas:

What about the Texas Two Step? Well, we had our own version of that. The bluff on which the site was situated was riddled with little crevices and cracks, which made for a great place for snakes to live and multiply. There was an overabundance of rattlesnakes and they enjoyed sunning themselves in places that an unwary airman might encounter, hence the unusual footwork that looked like a fancy dance step. But rattlers weren't the only menace in the creepy-crawly-flying menagerie that abounded on that Texas bluff. In the fall of 1958 there was a plague of locusts that was near biblical proportions, there were so many that vehicles on streets and highways had stopping problems. At night, the locusts would be attracted by the lights of the Air Police main gate and land on the road. As cars would come through they crushed the bugs which caused subsequent vehicles to come sliding through the security checkpoint. The diesel engines that powered the standby electrical generators would overheat because the insect bodies would plug the heat exchanger radiators. Turning off some of the outside lighting helped somewhat but just opening an outside door would attract dozens of the little critters into a building. A few months later it was a migration of millions of wooly caterpillars as they made their way south. We also had tarantulas, tarantula wasps, your standard red and black wasps, scorpions (big ones I might add) poisonous centipedes, copper head and coral snakes, and long before fire ants became a menace we had king size red ants. We all became adept at watching where we stepped or placed our hands when doing any kind of outdoor work.

Then there is the tale of the great rattlesnake hunt. An operator sitting at a radar scope raised an alarm, saying he had seen a snake poking its head up through an opening where cables exited the cable trough to the back of a scope. If you guessed that the radar maintenance troops were ordered to examine the troughs, you would be right. The great snake hunt was on. All of the building cable troughs were extensively explored, cables were poked and prodded, but no snake was found. Screen wire and Plaster-of-Paris was utilized to seal the building cable trough entry points. After it was all over there were a few comments from radar maintenance men concerning the ancestry of the scope dope. We also recommended that whenever this particular airman was on scope that it be classified Top Secret. If the Soviets knew they might try to attack us then. Lest you scoff at how serious the problems were at the site, a herpetologist from the Oklahoma City Zoo was commissioned to investigate and make recommendations about the collective menace on the site. Following several days surveying the bluff, and collecting at least 24-30 snakes of various sizes and species, he returned to Oklahoma City with his prized collection, and soon we were in receipt of his findings. The opening sentence has been stuck in my memory bank ever since, "Without a doubt the Killeen radar facility is infested with the most numerous and variety of venomous reptiles and insects I have ever seen." About the only recourse he could recommend was just to be extra careful. We were certainly that.

Later it came to our attention that the county had a bounty on rattlesnakes. Bring in a head and receive 50 cents. But we didn't find that out until we had killed about $50-60 worth.

Yeah, those were the good old days in the Texas hill country.

Jim Saunders:

We had a wasp infection on one of the radar towers at the 702nd AC & W Squadron, Savannah, GA in the late 60's. The air was full of wasps on both floors. They called in an exterminator from main base and he put those aerial insecticide bombs all over inside the tower. The next day we had a carpet of dead wasps everywhere. We swept them up and filled up a large cardboard box with them. I remember it being hard to breathe from the insecticide. I never could figure out why the radar didn't just fry the little buggars. I hate wasps.

Mike Gordon:

Desolated sites provided time to think and to do. One of the pleasures was the wildlife.

Going up Norad Road it was common to see black bears, wild turkeys and deer. I'm sure that mountain lions are around, too. Elk are a common sight out in the eastern portion of town.

Bobcats? How did I miss bobcats? I wouldn't have been surprised to see cougars up there. I do remember some of the troops that lived in the barracks finding a sick deer and bringing it into the barracks one night during the rainy season. Skunks were also a familiar smell. In one area I saw a deer that appeared to be somewhat street smart. He/she was looking around before it crossed the street. I saw another deer that was in a heavily traffic area that was in sheer panic.

Robert Chaney:

Oh yeah, the wildlife. When my draft board and recruiter helped me decide to join the USAF I visualized myself strutting around big fat shiny airplanes in a dark blue jumpsuit. Instead I ended up in slightly rumpled green fatigues standing next to radome looking out over some of the most breath taking real estate on this earth. And then there was the wildlife too! That recruiter did me a favor when he signed me up for AC&W.

Mike Gordon on the 776th Radar Squadron, Point Arena, California:

My wife and I arrived at Point Arena, California which had a population of about 300. The first drive up the mountain to the site on the curvy road was a harrowing experience and slightly unsettling. We had to drive through the clouds and when we got above them we arrived at the site 2600 feet above sea level. I signed in and found there were no quarters available for us so back down the curvy road and find a place to live. We got lucky and found a small cabin halfway down the mountain. The area abounded with wildlife, small California deer, wild boar, wild turkey and lots of rattlesnakes. It was normal to have deer grazing in our front yard or to see boar or wild turkey walking across the back clearing. It was the same in the site's family housing area. During deer season, the hunters stationed there would get up early, don their hunting clothes, chase the deer off their front lawn and go into the woods. They would hunt all day and not see a deer until they returned home and had to chase the deer off of the lawns of their government quarters in order to park the car. Hunting was not permitted on federal lands.

Lawrence Hamilton:

Indian Mountain was the worst. When I got there in Oct 53 it had just opened. I lived in an open bay above the orderly room. Operations and other things were on the 1st floor. Water was obtained by shoveling snow into 35 gallon garbage cans and melting the snow. This was used for cooking. Other water was hauled up from bottom camp in a water wagon behind a D8 Caterpillar tractor. I finally was able to move to the back side of the site to a Quonset hut where all of the other personnel lived. We walked the 1/2 mile to get to work when the weather was good. When it got bad we rode in a trailer pulled by that D8 Cat. Our latrine was a wood shack commonly called Outhouses. Under the hole was a 55 gallon drum with the top cut off. When it got full it was pulled out by that D8 cat and pushed over the side of the mountain. Thank God for Caterpillar.

I agree 100%. Life on a large base couldn't hold a candle to living at a radar site. In spite of the conveniences that a large base offers, you were still just a face and a name. On a radar site, you were a person - part of a much smaller group of people with a common focus. Friendships were stronger, work was better and so was the quality of life - even with the inconveniences and isolation from larger communities. There was a lot of pride in being part of that group.

Jim Sanders:

One day, about 2 or 3 in the Morning on the top site at Cottonwood, Idaho, one of the guys from radar maintenance came in and asked if we had a spoon. I said sure and went into the wire maintenance area and turn the lights on. I picked up a spoon and started to turn around when a mouse ran out from behind one of the jars on the shelf where I got the spoon. In reflex action I used that spoon to hit the mouse about 5 times in rapid succession and needless to say it killed the mouse. I turned around with the spoon in my hand and the guy asking for the spoon said "forget it." I can only think that he might have had a bowl of soup that he drank that night right from the bowl.

Guy Bierwirth Sr:

The mouse and the spoon reminded me of an incident at Hon Tre Island, Viet-Nam topside chow hall. The chow hall was just a tent with plywood floors. While eating dinner (for lack of a derogatory term for the slop), an Army guy next to me asked if I had a knife. I handed him my folding knife which he opened and reached under the table towards his boot. We heard a sound like a screech. He brought my knife back up, wiped off the blade on his fatigue pants and handed it back to me. He had just decapitated either a large mouse or a small rat that he was holding prisoner by its tail with his boot. Suddenly, the slop looked even worse. There were so many rats there that every night we had a rat patrol. We would send out a detail to go around the site topside and collect all dead rats and throw them in a half 50 gallon drum, and mix with diesel fuel and burn them. Then we had to put out fresh poison and start all over again.

Lee Dixon:

Here are a few short notes about rats' topside at Hon Tre Viet-Nam.

When I first got there guys from the Army HAWK Battery would catch rats and hang them in front of the acquisition radar until they exploded. There was a 1948 shelter being used as the operations office. Late at night when it was slow, we would go into the office and sit very, very quietly in a circle around a tin of peanut butter. In a little while a rat would come out and we'd take all kinds of implements (rolled up paper, etc) trying to beat it to death. One night I was on scope drinking coffee and I had some traffic check in. I set the coffee down to handle the traffic. When I finished with the traffic I reached for my cup. When I looked down, there was a rat the size of a small cat drinking out of my cup. I screamed and one of the NCO's got a rat trap and used peanut butter. He set it in surveillance near the edge of the tent. Within the hour it snapped and we had him. The rat then pulled the trap almost to the plotting board before giving up the ghost. It was a big male that could almost have bred with a small dog. The food from the mess hall that wasn't eaten went over the side of the mountain toward Nha Trang. At night you could shine a light down there and see many, many sets of eyes.

JOKESTERS AND ANTICS

Leave it to a GI to come up with pranks, jokes and ways to initiate a newly arrived Airman or Officer to a radar site, especially a site that has a specific tour length because of its isolated location. It is almost a responsibility for the man who has been there the longest, called the "Short-Timer" to taunt the new arrival because he has more time to serve on the site than anyone else. Here are a few:

Ray Buda, Chief Master Sergeant, Retired:

When a new 2nd Lt arrived at Cross City AFS Florida and was assigned as Ops Officer. Our resident jokester Joe Alsup, was filling is as Ops Office Clerk, informed the brand new Lt that the coat rack by the office door was reserved for enlisted and the one at the far end of the hall was for Officers, the Lieutenant used that coat rack for well over 2 months until the Commander came in one day and asked why he was walking all the way down the hall just to hang up his hat. The Lt had a sense of humor and Joe survived to joke again.

We had a fun trick that we did when we got new operations folks from Keesler when I was stationed at Wallace Air Station in the Philippines. One of the more experienced operators would ask me to set up targets on the system. He would set up those targets to be accepted on the radar console where the new man was stationed. I would send up 12 aircraft coming to the site and the poor young person would get all excited about these approaching aircraft to the point where he was almost shouting at the controllers to do something, send up the interceptors. Of course the controllers knew what was happening and would not react. It was funny of course when it was revealed that we were not being attacked and it did have a good result as well. It calmed down the new operators and showed us how the guy might act in a real emergency.

Robert Wilder:

This reminds me of the time while in base comm at Eglin AFB Florida, of a joke we always pulled on any new Second Lieutenant. If maintenance control got a call from Field 5 as a trouble call, we always asked the new lieutenant if he would like to go along. Field 5 was near the Army Ranger training camp so we would call them and see if the mess sergeant could have lunch ready for us on our return. We would stop and have a great meal and start back to our main base at Eglin. We would steer the conversation to the fine lunch and inevitably, the lieutenant would comment on how good the chicken was. That was when we told him the chicken was really rattlesnake and that every other item on the menu came from the wilds of Florida. Needless to say we had to stop while the poor "butter bar" left his lunch by the side of the road.

Joe Baude:

The site was "P" Mountain, a few miles from Thule AFB in Greenland. The unit was the 931st AC&W Squadron. The year was 1953. We had learned the fundamentals of electricity, electronics and systems. We were now learning to maintain systems and our life with strangers and friends living in relative remote site isolation environments.

Our First Sergeant was Joseph A. Yulliano. Sometimes a First Sergeant acts routinely in ways that irritated the "troops". Every order he issued was "posted" on bulletin boards with his initials "JAY". He was an egotistical command authority freak, conscientious and over-zealous. That is probably how he had been taught. It was JAY ordered this and JAY ordered that, day after day and the troops started to rebel as best they could. It started with small pranks. Like a pie would be taken from the mess hall and in the empty pie tray was a note with the words "OK JAY". Then in the latrine, a number of rolls of toilet paper had every sheet imprinted with "OK JAY". Then there were JAY orders posted with big red X's of cancellation notice with a signing of "OK JAY".

Someone went down to Thule base, visited the library and signed up for a weekly mailing of a Fundamentals of Electricity correspondence course to be mailed to JAY on P Mountain. When JAY questioned the troops as to why he was getting this material, a note was posted on the bulletin board that noted "the Air Force recognizes the weakness of it's leaders and automatically enrolls them in courses that will help them learn what they need to know" – the note was signed "OK JAY".

The ultimate prank was an elaborate set up. On Saturday nights, the NCO and the Airman's clubs were open until one AM, sometimes later. The laundry cart was big enough to contain a big trooper. A few airmen went to JAY's room at midnight, woke him up and told him a few guys wanted to apologize to him in front of the guys at the NCO club for their "OK JAY" pranks. They convinced JAY of their sincerity and got JAY to get dressed. They told JAY that they would TAXI him to the NCO club down the hallways to show that he was being respected for his command authority. They did a good job because they got JAY to get into the laundry cart. As the cart approached the last hallway on a downhill run to the NCO club, the troops turned the corner and pushed the cart to run free, down the hall. The hallway was lined with troops holding signs with "OK JAY" on them and cheered JAY on his high speed, downhill, taxi run to the NCO club. JAY did change his command style a bit after that night.

James Joslin:

We had a guy in on Temporary Duty (TDY) that was causing trouble for the regulars in our Barracks so someone placed trash can full of water leaning against his door. The bummer was that he was on the second floor. When he opened his door the can spilled onto the floor and into his room. Then the water soaked through the floor into the room below.

Jan C. Hoffman Sr.:

When I was at Benton AFS PA, they were having trouble with the trace on one scope looping. All the radar troops were stumped and the Chief of Maintenance called to the GATR site and asked for two of us radio troops to come back to the main site and look and see what we could find. We ran through all the alignments and checks the radar troops did and tried a couple of things we thought of with no luck. Ops was changing shifts and one of the troops coming in pulled out a piece of rabbit fur and yelled across the dias to another troop " come over here and watch this" while he rubbed the fur across the plastic protector on the screen face causing a severe case of static electricity to effect the trace. We then knew what was causing the problem and advised the Radar troops. The young man in ops saw the error of his ways when somehow a capacitor seemed to discharge in his vicinity several times over a couple of weeks and he received the jolt of electricity. He also learned why you don't make extra work for maintenance.

Jay Tumey:

Upon my return from leave to Cape Romanzof, Alaska I was greeted by several troops and immediately escorted to the club for a welcome back. After several drinks I declared it was time for me to retire to my room. I walked down the hallway where I thought my room was located. It wasn't there. I thought I might be in the wrong barracks, but I checked and I was where I was supposed to be but my room wasn't. I took another route going down another hall but it was not there either. As I looked back where I had just been, I noticed our site carpenter standing in the hall laughing. He went to where I thought my room should be located and as a crowd gathered he removed the plywood sheet from my door way, complete with tape and paint job.

Everyone had a good laugh on me looking for my lost room.

Bruce Shymanski: I have a couple stories from my days at the 874th ACWRON, Inoges, Spain:

We had two radar towers which were connected to other areas via a network of enclosed walkways.

The door to the search antenna tower, where the maintenance crews hung out, was controlled by an electric keypad. The keypad system was hardwired and the combination changed by moving the wires around – high tech for the times. We would change the combination when we came on duty to keep officers out. I was on the American crew and we relieved a Spanish crew. We would sometimes arrive at the tech site early and climb a rope, which we had earlier strung out, to the second floor of the tower. We would then go to the area above the entrance door and peer down through a hole in the metal floor. We could see the door and the area around it. Being the technical wizards we were, we had earlier run a couple wires up to the 2nd floor so we could open the door remotely.

From the second floor, we would unlock the door, which made a racket when the solenoid kicked in. The Spanish AF crew would wait, expecting someone to walk through the door. When they heard no further activity, they would check the door and then look down the hallway. Of course there was no one there. They would return to their chairs. We'd give them a couple minutes and unlock the door again. This time, they would get to the door a little faster. Again, no one was there. Now they were perplexed. They would hang out by the door waiting for it to happen again. Of course it didn't happen because we were watching. Once they gave up and returned to their seats and we would do it again. This time they would be up muy pronto and run down the walkway looking for the bad guys. It drove them nuts. After a few times of this, we'd go back out the window, around the building and come in the door. The Spanish AF guys would be anxious to tell us to watch that door because something was going on.

We would repeat this exercise every couple months.

After our shift, the older guys and married guys all went home to Zaragoza every night leaving a hundred young guys loose in the cantonment area. You would never see an officer around at night.

There was no TV or radio in those days but we had a great NCO club with fifteen cent drinks. If we couldn't afford that, we'd walk down to the pueblo where we could get a glass of wine for a couple pennies, a whole bottle for a dime. Since we were, comparatively speaking, big-spenders, the three bars about each the size of a bedroom, would stay open as long as we stuck around.

On Friday nights the club had nickel beer at happy hour. We couldn't wait to get off the hill for that. Often, if we had some money, we ran down tripping and tumbling along the way, arriving bloodied and battered. Being the clever guys we were, we would pre-purchase a case from the Spanish bartender and ask him to keep it cold for us.

I went back a few years ago and had a glass of that wine. I couldn't finish it, it was that bad. I guess my tastes have changed along with my age. I think of those days as "Young men without adult supervision".

Larry Knaff:

Some may not call this a practical joke since it borders on insubordination, but it was funny and was a great morale builder. I arrived at Cut Bank, Montana in January 1956. Our base commander at the time was Captain Robert Wilson. He was a great guy, but he didn't run a very tight ship and there was very little military decorum. I recall several memos addressing this deficiency being posted on the orderly room bulletin board. One memo deplored the lack of saluting. Captain Wilson stated that saluting was a matter of respect and that he as Base Commander deserved at least one salute each day. Another memo targeted the lack of rank insignia on the rag-tag fatigues that we called uniforms. The troops loved the laid-back atmosphere under Captain Wilson. The morale was high and we got the job done.

In August 1956, Major Richard Atchison, Jr. replaced Captain Wilson as Base Commander. Major Atchison was appalled at the "Camp Swampy" atmosphere and laid down a number of new rules in a memo called "The Honeymoon Is Over." Needless to say, military decorum improved and we started to take on the appearance of a military operation. But the change was a big shock to all of us and morale declined.

In early 1957, a bogus memo appeared on the orderly room bulletin board entitled "The Honeymoon Is Over - Part II". This memo looked very official and had the signature of Major Atchison. It looked official but the document was obviously a parody on the original memo. Major Atchison was furious, locked the gate, and said nobody would be leaving the base until the perpetrator of the bogus memo came forward. About a day later, the restriction was lifted with little fanfare and things got back to normal. Actually, morale was greatly improved as a result of the humor and everyone was talking about it.

Of course, everyone was interested in what happened and who wrote the memo. Officially nothing much was said, but eventually the truth came out. A young lieutenant supposedly wrote the memo. This was a great surprise. Nobody expected that the perpetrator would be identified as an officer. And the lieutenant in question was a person that you would least suspect of doing something like this. He was, as I remember, a graduate of the Citadel and had exceptional military bearing. He was responsible for organizing a Squadron Drill Team and served as its commander. The whole incident was swept under the carpet and the lieutenant remained in good standing at the base.

Bob Wildrick:

Back in the late 60's I was working in switch engineering at Stromberg Carlson. As a desk jockey all of us had watts lines switched to our desks on the day shift so we could make the necessary telephone calls free to other locations. Occasionally the operators would forget to switch off these lines at the end of shift.

One of the engineers had a particularly difficult job which required him to use a watts line to call another location. This engineer was a taco short of a combo platter. The switch engineer called out on one of the watts lines and the call came into the engineer's desk. The switch engineer pretended to be the customer and started to give the engineer all kind of grief over how he was designing the power distribution for the particular office. I kept watch over this whole thing but couldn't laugh as it was going on. It was even harder to keep from laughing when the engineer came back to the switch engineer's desk telling him all about the call with the "customer." He never did figure out that the call came from right there in the office.

Jim Sanders:

In the mid fifties, we were making $76 per month and the government did us a huge favor by putting us in the Social Security Program taking even more money from our dwindling pay. We were lucky to have enough to buy a beer after paying income tax, deductions for the old soldiers' home and for the bonds we were encouraged to purchase. True it only cost 5 cents to get a soda from those old soda machines. We had to really pinch pennies to make our money last to the end of the month. We also looked to take advantage of any weakness we could find. The model of soda dispensers we had at Cheyenne were easy to beat. Put your nickel in push the lever very slowly till you heard it dispense the bottle then back off just a tad and do it again. A few of us emptied the whole machine by only putting in one nickel coin. Then we sat around and drank every bottle. We must have each had a case. The machine only had Doctor Pepper in it and to this day I cannot drink Doctor Pepper. It is one thing to sit around and drink a case of beer. You get drunk and you have a hangover for a day. But to sit around and drink a case of doctor pepper makes you sick and it lasts for days.

There was another kind of drink machine where you opened the top and drinks were hanging by the glass bottlenecks. The get a bottle, you put your money in and slid the soda to the left and removed the bottle. Uncap it and drink. The guys quickly learned that by removing the cap from the bottle while it was still hanging in the machine you could use a straw and have a drink without paying. The machine would end up with empty bottles still in the secure racks in the machines.

I remember once at Fort Campbell Kentucky I wanted a cola. I put in 5 cents and it returned four nickels and a bottle of coke. I thought maybe I had put in a quarter so I tried it again making certain I inserted a nickel. Again I got the soda and four nickels. I quit when the machine ran out of nickels.

I also had fun with the early cup machines. You could allow it to start the drink pour then unplug the machine. It would continue to pour indefinitely. A person could get several cups filled if they were fast enough to make the switch and they had a bunch of cups available.

I always wanted to write to these soda machine makers and recommend they place any new machines they developed into a military barracks for security testing. If there was a way to cheat the machines, a GI, earning just enough money to not have any fun, would find a way to get the product free.

Donnie L Shahan:

In 1964, I was stationed at Seymore Johnson AFB, SC. We had a man in our detachment that never seemed to think logically. He would say some of the most off the wall things and do the odd things we couldn't figure the reason for.

One day there were five of us in a work detail. The Staff Sergeant in charge of the detail told the odd fellow that he would buy a Coke for us all if he would walk over to the recreation room and get them. The Sergeant handed him five dimes and he only brought back four Cokes and handed one out to us but none for the Sergeant The Sergeant asked where his Coke was and the odd fellow replied, "Sorry, Sarge, the machine took your dime."

We stood there in awe. He was never even aware. The four of us then kicked in and between us we had another dime. We sent the oddball back to get a fifth bottle.

Bill Lechien:

When I was on Fire Island, Alaska in '66 it was considered a remote assignment. At that time it was Alaskan Air Command (AAC) policy, if you were on a remote site, no leaves, 3-day passes etc., unless it was a bona fide emergency. The commander adhered to that policy to the letter, even though we had supply choppers coming to the island Monday through Friday, weather permitting. The worst part about being on Fire Island was that on a clear night you could look across the water and see the lights of Anchorage. The city was so near and yet so far away. Fire Island had a change in command as is normal on remote sites. The new commander, also a Major as was the old Commander, must have looked across the water on his first night there. The next morning he held a very impromptu Commanders Call. He took the stage and made one simple statement that lifted the morale of the entire site. He said, "Gentlemen, its AAC policy that since you are on a remote site, you cannot have a leave or three-day pass unless it is an emergency. But, I can damn well send you on a three-day permissive TDY over your break."

He left it up to the section supervisor's and first sergeant to schedule those permissive temporary duties off the Island. We had to go on our first day of break and be back on our last day of break depending upon the flight schedule. Most of the time we only had about a day and a half on the mainland but that was long enough to take care of what needed to be done. At least it was a break from the Island.

Jerry Frey now in the Ozarks:

I remember another time when a couple of the chopper pilots were flying down to Seward Alaska to get some stick time in and to go fishing. They made arrangements to stop at the Fire Island and pick up whoever wanted to ride along with the commander's okay. We just finished a midnight shift and got to the strip just in time for the all day ride. Sure was great that the pilots thought enough of us to offer us the ride.

Jim Sanders, with another story:

For sure guys we got away with some things. At Cottonwood Idaho and I would guess it is similar at most Radar Sites, the top site was accessible from the bottom site by a one lane road. It was about 12 feet wide with snow poles marking the outside edge. To travel up or down the road you had to request permission which was always granted unless there was someone traveling up or down at the time.

Our transportation was a six by six, two ton truck which we used to make the crew changes. Between the top site and the lower site we had three emergency phones so that if you broke down you could get help.

Our Air Police guys had got out their math book and constructed a table that more or less said if you left the top site at such and such a time and it took less than eight minutes and 30 seconds to get to the bottom you were speeding and they would give you a ticket. After several of our drivers getting this calculated speeding ticket we devised a plan to cause the questioning of the mathematic ability of the Police scheme. The plan involved sneaking down the road without asking for permission to use the road. Make the phone call at that last emergency phone and then put the truck in a very low gear and tear down the road. There at the bottom of the hill was the Air Policeman with eyes as big as saucers as this was off the chart which was now discredited. Everyone in on it got a good chuckle out of that. We never told the Air Police what we had done but their chart went into the wastebasket.

Robert Chaney:

On the road from the town of Galena, Alaska to Campion AFS there were either three or four check points from where the drivers had to call back to Campion AFS Operations that you had crossed. Each vehicle had a radio like the old CB's in it. As you passed the check point the driver would typically say, "Mail truck at the dip" or "mail truck at checkpoint three." Operations would repeat the location and log it in the event they did not get a call in a reasonable period of time, they could send out some help. Sometimes some joker from operations would request we stop and get him fries and a coke and we all would all laugh like we never heard the request before, which we did on almost every trip.

Mike DeRoss:

Yes, the daily mail run from Campion to Galena, Alaska I must have made that damn run about 300 times. As soon as you got in the truck and were leaving the motor pool it was "vehicle 73B4094 leaving site." Just across beaver creek was checkpoint Alpha, Bravo was down the road a bit, Charlie was about a mile before Galena, then you had to call in "on the dike" and finally that you were in Galena. The problem with calling the checkpoints is that the Motor Pool Chief and the Commander could tell how fast you were going if they were listening to the radio. The speed limit was 25 or 30 MPH depending on the place. My best time was 11 min for the nine miles from Campion to Galena. It was always faster in the winter on the snow than on the gravel, potholed road. I learned quickly, well not so quickly as it did take a couple of butt chewings, that there was a way to fool them. If you timed the calls, no matter where you were, everything looked fine. Many times I was in Galena having a coke when I called and said I was still on the road. I figured that if I put the truck in the ditch they would find me soon enough.

Pete Callamaras:

Years after I left the ADC world, I was stationed in Germany. One afternoon as I was leaving our building, the Captain in charge of another shop was coming up the walk between shops with two bags of groceries in her arms. I saluted, and, as she started to try and figure out what to do, I laughingly said "That's okay Ma'am, you can owe me one." She chuckled and we both kept going. About a week or so later, I was taking some equipment between some buildings with my hands full. She must have seen me coming because she came out of her building, pops to and salutes me. Then she laughed and said "Okay – we're even."

Joe Sargent:

As for the saluting, I remember when some "Good Old Boy" NCO made Warrant Officer we intentionally got in front of him to salute him and force him to return the salute. We had a game we pulled on some of these new Warrant Officers. We would see him coming down the street and we would position ourselves to salute him. After he returned it and went past us, we would run around building to get in front of him again. It was like there was one guy after another, wherever he turned, it was salute after salute till he took off running to get away. Another time a few others and I were really just walking down the street. The Warrant Officer turned a corner and ran into smack into us. We saluted him not knowing what was going on previously, he threw us a haphazard salute accompanied by some profanity.

Ray Garber:

At Benton AFS, Pennsylvania during the early 60s, we had radiation badges (little blue plastic clip-on jobs with some sort of film inside) that had our ID impressed on it. We were to wear them every time we went out to the towers. We often wondered if anyone ever checked them, so one time while out in the search tower, we decided to put one under the lead cover on the magnetron. I had forgotten about it until we were out there again for some checks. I took it out, and pinned it on and never gave it another thought. Routinely, we exchanged the old badge for a fresh one, and after such a change, I was called over to the medic's office. I was asked how I had been feeling lately, any nausea, etc. I was dumbfounded, as I had forgotten all about putting that badge under that cap. When he told me that my badge indicated that I had been subjected to a heavy amount of radiation, I became very dumb, very quickly. They later concluded that it must have been a fluke, as I showed no ill effects. Didn't try that trick again!

Roy Killion:
At Miyako Jima, my friend Pete Godlewski and I decided to put some small florescent lamps taped to the outside of the search radome in the shape of a Christmas tree. We thought it looked real neat as they would light up as the search antenna rotated by them. We got a call from the Commander and ordered to remove them that night as the Ruyakan local nationals working on the site had left their work areas (Security, Power Production, Club employees) because they thought that the site was haunted! Nearly every year I get to thinking about it and can't keep from laughing.

William Elliot:

Around Christmas of '62 at Wheelus AFB, Tripoli Libya, we made a star out of 4 ft florescent bulbs and mounted it on the back side of the search antenna. When the height finder wasn't being used by Operations, we pointed it at the search tower. We got a nice flashing star with every nod. That lasted about a day before we were made to take it down. I guess the Muslims didn't like it and we were guests in their country.

Vic Tanner:

Another story someone told me concerned an argument between the hilltop cook and a radar technician. It seems the radar tech claimed he could cook a steak faster than the cook. I was told they stopped the height finder, fired up the transmitters and put a steak on a long stick and nuked it in front of the feed horn. The Technician won.

We would also throw up a flashbulb in the height finder radome in front of the antenna and watch them go off.

There were rumors that by sitting on the feed horn it would make you sterile and that many of the married men had taken up the practice. I think that receiving that much radiation would be lethal.

Ken Jones:

I was assigned to the 914th Radar Sq in Armstrong, Ontario, Canada from 1961-62, and if you had your family with you, it was considered to be an accompanied tour, but if unaccompanied, as most of us were, it was a remote tour. Only about half of the officers assigned had their ladies with them. If you loved to hunt and fish, as I did, however, you were in Hog Heaven. We had several fishing camps with boats we could sign out. We could only get to some of them in four wheel drive vehicles so special services had one of them we could sign out as well. It was the only place I ever heard of that when you arrived you got a free Canadian fishing license as part of the in-processing.

David E. Casteel:

Opheim AFS was so far north in Montana that at retreat they played O' CANADA. When anyone got a three day

204

pass they went to Yak Montana, (population really small) just to see the bright lights and enjoy the night life.

Although Mt. Hebo AFS, Oregon was out in the sticks, it was only considered "semi-isolated" because there were 27 units of family housing allowing some guys to be accompanied. Those married folks who did not get military housing often found quarters in the nearby village. I was told that they were substandard and expensive. Myself, I was single and lived as the only occupant of the Bachelor Officer Quarters. I spent most of my time on the Hill and when not at work I freely associated with the other station residents, at work or play. I went down to the Operations area quite often in the early morning hours, just to chat and see what was going on. I also saw people in the Recreation Building and the NCO Club when I was an invited guest. I never did enter the enlisted Barracks. I felt that would have been an invasion of their privacy. I don't think my activities would have been considered "fraternization" because there were no strong friendships developed; it was just the way I maintained communication with others and kept from going stir-crazy.

Tom Clason reported:

I heard some stories about guys' playing cards while on duty or watching television. While assigned to a site in British Columbia, Canada, I never saw a TV set in ops. I remember a lot of "dozing" in the break room and I remember us giving a guy a hotfoot in there. There was also an airman who fell asleep while on the scope. He lost a stripe as punishment. I remember the scope man and the plotter pulling some "pranks" drawing things on the board to see if anyone noticed. I also remember crashing a garbage can like cymbals behind a guy on scope. There were times when the radar was down and then only two of us would be on duty. Then we would spend all night talking and telling lies to Canadian WAFs in Vancouver. There was no reason to play cards, talking to the girls was much more interesting to us.

Living on the economy is quite hard on new GI's especially if you are among the lower grades or ranks. I worked as an Airman Second at the radar site outside of Petaluma, California. I got married and fortunately for us, I got promoted to Airman First in just a few months. The assignment was a good one for a guy with a new wife because we worked six days and off four. When off duty, I worked at least two other part time jobs. We had to rent a place to live in Petaluma and I must say that the converted chicken coop we found was clean and almost looked like a real house. Ours had one bedroom, one bath and a very small kitchen, but we made due. But then as newly-weds, how much room do you really need?

Gary J. Wozniak:

In all the nominations and discussions about the most isolated site in the continental United States (CONUS) I have not seen a comment about Finland, MN. Some sites had top camp here, housing there, admin over there. At Finland everything was at the top of the hill, radar, ops, admin, housing, everything. There were many times when the site was cut-off from all below because of snow blocking the access road. This meant everyone at the Site, active duty GI's and dependents alike were there for the duration. There were times when this lasted a few days. Duluth was 60 or so miles away. Silver Bay was fairly close, about 15 miles. It was an inactive reserve iron mining town and there was nothing to speak of there. Two Harbors was 40 miles away and again not much there. All that isolation, but I loved it there. Kids did too. I can't count the "forts" in the woods they made over the years. I guess it was far enough North where I never saw a tick on the dog or the kids. Calumet had the snow too. Some snow forts were 8 feet high after the blower came by and made a nice base mound to start off construction.

Patrick Donahue:

In the fifties officer and enlisted latrines were segregated. I remember when I was stationed at the Pentagon I had a bad case of the diarrhea and I knew the enlisted latrine was quite away from where I was and I would never make it, but the officer latrine was real close so I took a chance and when in.

No one was in there so I took a stall and proceeded to relieve myself of that nasty diarrhea. No sooner had I passed that and some gas as well, I heard someone come in. The first thing he said was "Whew, What the Hell" I said nothing, hoping he did not ask who was in the stall making that nasty smell.

He left quickly and I did too. I often wondered if I got caught using the officer latrine would they have punished me.

Jack Sheehan:

32 miles from downtown Las Vegas Nevada sits a mountain called Angels Peak. Atop Angels Peak sat the 865th AC & W Squadron. The elevation was 8,868 feet above sea level and our barracks and NCO club were just 500 feet lower. That was high enough for me. Great

place for beer drinking, one beer and you were on your way, but it was a bad place for playing basketball. I'm not sure if any sites were at a higher altitude.

An unknown airman submitted this story:

I pulled Kitchen Police (KP) in an Army mess hall one weekend day every month for the three years I was assigned to the 662nd Radar Squadron located on an Army Post, Oakdale Army Installation, Pennsylvania from 1963 to 1966.

I found that Pots and Pans was not the worst job in the mess hall. It was a hard job, for sure. But the worst job was the milkman who had to lug all those stainless steel milk containers delivered weighing in at around 50 pounds each. Each can had to be rotated inside the walk in chill box so the oldest milk would be used first.
Another job I did not like was on the dining room crew. They had to clean the kitchen floor and the dining room floor between meals, waxing buffing and such. They never got a break.

I always came in last to KP so I would be assigned pots and pans. One of the benefits of the job was that when I had to go to the latrine I just went as often as I needed to without asking permission. If I didn't clean the sheet pans well enough from the breakfast bacon, as well as the large pots, the cooks would have to use them anyway. If they did, possibly 500 diners were going to get the "G.I's" (diarrhea) so they relied on me to get the pots and pans good and clean and I tried to not disappoint them. Because of that no one messed with me.

There was always a KP pusher in the Army mess hall. A "pusher" takes great delight in making sure that the five soldiers and two flyboys on KP never get to sit down until it was their time to eat their meal before the rest of the crowd came in for chow.

Of course, as I did Pots and Pans all the time I did get good at it and soon developed a routine. I actually became so efficient I had time when I had nothing to do. That is not allowed in the military especially if you are a trainee. So, I was assigned to also operate the "clipper" running the trays and bowls through the dishwasher. The bad thing about the clipper is that some wise-ass army draftee would take soup, not eat it, and then sling his tray into the clipper causing the soup to splash up on the KP's fatigues. Your front was usually a wet mess by the end of the day. All the grease and hot water from my jobs as a KP ruined a set of fatigues and a pair of brogans. Rather that ruin other clothes, I made them my KP clothes and I kept them in the trunk of my car. Whenever I was assigned KP I put on my KP uniform. The head cook could not figure out how I got so nasty looking in such a short time. If he got there at the same time as I did he would have known.

Jay C Phillips CMSGT (Ret):

I can clearly recall the "Thumm-Thumm-Thumm" of six pusher engines on each of many B-36 aircraft as they flew low over my family home in Irving, Texas. Their distinctive pulsating earth shaking sound rattled the windows in our house and vibrated our bodies. The aircraft were returning from a fly over at Washington DC, celebrating the establishment of the landmark 1947 National Security Act and the creation of the USAF as a separate service. That September 17th day was also the beginning of my fascination with the Air Force that continues into the 21st century.

After graduation from high school and, with the Korean conflict ongoing, I was leery of being drafted. During the previous three years I had buffed and polished all the brass buttons and belt buckles I cared to. Knowing that the AF uniform had oxidized silver colored buttons, lapel insignia and belt buckles which didn't need polishing, helped convince me that the AF was preferable over the Army.

On April 13, 1954, after being pronounced hale and hearty at the Kansas City induction center I was soon on a train with several other enlistees on the way to Lackland AFB, San Antonio, TX. We arrived about 24 hours after departing KC. There we boarded an AF bus and enjoyed a last moment of peace before being politely asked to leave the bus at the base reception center. Perhaps politely wasn't completely true. Apparently the Sergeant that boarded the bus thought we all had a hearing problem, as the request was delivered in a thunderous voice that could have pierced plate steel.

Without exaggerating I can state that I was as well versed in small unit drill and ceremony as most any Basic Military Training Instructor. Moreover I knew the manual of arms for rifle, pistol, and saber. In addition I was an expert rifle shot, having fired in several competitions. I had soaked up everything my JROTC instructors and CAP trainers had to offer. This was to be, sort of my undoing.

Assigned to the 3725th Basic Military Training Squadron (BMTS), Flight 412, I was appointed Student Flight Leader. Suffice to say, that put me in charge of the flight whenever the Flight TI was away. I marched the flight to and from many training sessions and was in charge of the flight and barracks when the TI was absent. I was always excused from any details. This did not endear me to many of my fellow recruits, and was my first experience with RHIP, rank having its privileges.

Then came the day the flight was scheduled to report to the personnel center to process for assignment following basic training. My TI instructed me to remain in the barracks as my assignment was already taken care of. He told me that I was to become a Training Instructor and would remain assigned to the 3725th BMTS. Our training was coming to an end with a bivouac encampment and live firing of the .30 caliber M1A1 Carbine. At last after the 90 days of basic training the new airmen were scattered to the four winds of Air Force Technical School training. All but me, I was busy sewing on the single chevron of an Airman Third Class and soon found myself being given on the job training as a Tactical Instructor while waiting for the next TI school. A single stripe was low on the totem pole of rank, but the distinctive TI epaulets and white pistol belt denoted one as a TI. Those separated the have from the have-nots quite nicely. In the Air Training Command being assigned as permanent party garnered a few perks, chief among them was being eligible for base housing. Following a short leave my wife joined me in setting up housekeeping in a two storey on base housing unit. Life was definitely looking up.

With my graduation from LAFB Instructor School and a certificate of completion specializing in "Tactical Curriculum, General Instructor" I was soon assigned a barracks and went to the base reception center to pickup a flight of "rainbows." The days and months that followed were a blur of days with long hours, as the challenge of converting a gaggle of 60-70 young men into a smoothly functioning flight of airmen.

February 1955 brought a welcome surprise, promotion to Airman second Class. I now had an assistant. Following the cessation of Korean hostilities caused a scramble for ways to reduce AF basic training costs. One of those was elimination of weapons qualification and bivouac training. Familiarization with explosives and gas mask training was also cut. To further speed up the training pipeline two phases of basic training were established; Phase I was six weeks basic training at one of the three basic training centers(Lackland, Parks or Sampson AFB), followed by Phase II consisting of six weeks training at an Airman's Technical School. This change in basic training would further enhance the budget picture and shorten the pipeline time to put airmen into their respective career fields.

8 February 1956 came with Special Orders assigning me to the Training Wing Lowry AFB, Denver, Colorado training students to be Munitions and Weapons handlers. We soon found ourselves conducting barracks inspections and supervising work details.

Barracks and personal inspections were held every Saturday morning. Failure to pass muster at inspection would prevent an Airman obtaining a week-end Class-A Pass.

My immediate supervisor was a tough but fair minded MSgt. He had entered service at the beginning of America's entry in to WW II and when the war ended; he was a Major who had survived a European combat tour piloting a B-25 medium bomber. To remain on active duty following the end of the war he accepted MSgt rank, with an AF Reserve Major's commission.

Today's Airman would find it difficult to cope with barracks life as the 50's students did. Barracks life was Spartan; double bunks, a foot locker and a shelf fixed to the wall with a clothes rod were all the luxuries one had. The BX of the time mainly stocked necessities and few luxury items. Few students owned a vehicle. Firstly there were very limited parking spaces and, secondly the pay scale for junior ranking airmen was quite low. Paydays were once a month.

June of 1956 brought promotion to Airman First Class and upgraded my Primary Air Force job description from apprentice to skilled Training Instructor.

Things rocked along about the same for the next several months. Some of the interesting occurrences;

An airman was disciplined for riding his motorcycle thru his barracks one Saturday night.

A base dining hall cook was caught leaving the base with a bag full of T-Bone steaks and several pounds of butter. It seems he had been augmenting his home food stock for some time. His diet underwent a severe change when he was sent to the stockade.

I caged some more aircraft rides. The Squadron Adjutant would take me along when he flew his required monthly four hours to maintain his rated status. He enjoyed doing acrobatics' in an AT-6 Texan trainer. I didn't always enjoy them as much as he seemed to.

1 September 1957 was another milestone. I was promoted to Staff Sergeant. Not bad, three years four months, and three weeks to Staff Sergeant rank.

Seventeen days later I was assigned to Keesler AFB, Biloxi, Mississippi for the purpose of attending the Airborne Radio Operator course. At last, I was on my way, only a few more months and I would be graduated from my technical school and finally reassigned out of the Air Training Command.

But no, another surprise awaited me at Keesler AFB. My course assignment had been cancelled while traveling to Keesler AFB. I and several other students would be trained as Aircraft Control & Warning Radar Repairman, a twenty week training course. With E-5 rank, I was now entitled to have my household goods shipped from Joplin MO to KTTC but it took until January 1958 for that to happen. It was a belated Merry Christmas when my family and I were once more reunited.

I then looked forward to completing technical training school and finding out where my assignment in my new career field would be.

Came the day of graduation and we received orders transferring all of us to various radar sites. I read my orders and did a double take. I uttered a few choice words and called the idiots at personnel some unflattering names. My orders read that I was being assigned to an Air Training Command Base. Was I never going to escape the clutches of the ATC?

Henry Powell Jr:

When assigned to Lake Geneva, Wisconsin we got along with the local population very well. They were a lot of great folks. It was home away from home for a lot of young airmen from the 755th. Beer was ten cents and if you were short on cash your credit was good. My favorite hangout and bar was Pat Granahan's. I left in December of '53 and went back 20 years later when I retired. Vie Granahan called me by name when I entered as if I never been gone.

Fred Kemp:

Since we are on the subject of favorite bars, those of us that spent any time on the Airborne Command Post at Offutt AFB, NE will remember the Golden Horseshoe. This was the favorite for all of the crews and we spent many an hour there discussing flights, mistakes and even accomplishments. We even held a reunion there many years ago. Sorry to see that it is gone now. There are many happy memories of the place.

Chuck Miller:

At the 922nd AC & W, Goose Bay Labrador they had a drink in the Officers Club called an Afterburner. It was cognac poured into a narrow mouth cordial glass then lit on fire and drunk. Only seldom did anyone actually get burned from this practice. One evening, Major Thayer, the site commander, and a few others were in the club and decided to have an afterburner. There was no cognac left and there were no narrow mouth cordial glasses. (I understand that there was an occasional throwing of glasses against the wall. However my lawyer has suggested that I not discuss any involvement in throwing glasses or reveal any knowledge that I have of anyone else throwing glasses.) This one night there were some burns suffered among the officer corps. My recollection of the incident goes like this.

I was playing pool in an adjoining room and did not witness the actual event. It is my understanding that the group substituted Lemon Hart 151 proof rum for the cognac, and wide mouth shot glasses for the narrow mouth cordial glasses. It was a recipe for disaster. They all together tried to toss the burning liquid "down the hatch." It spilled out both sides of their mouths and were burned on both sides of their faces from the corners of their mouths down to their chins. I saw them when they came out of the club and they were all in pain. All of them went to the medic. I heard that he put some salve on it and gave them a safety lecture. The burns created a very raw area on their faces where they could not shave for a number of days. Thus we had three or four officers walking around the squadron with Fu Manchu moustaches. Either the medic or maybe the fire marshal declared that there would be no more drinking afterburners, but the site commander quickly countermanded that directive. I don't remember any lasting scars on their faces, but that would have been possible.

I also recall being asked to go to the Officers Club at the main base and to get a box of the narrow mouth cordial glasses to replace the ones that needed replacing at the site. When I took the box of them over to supply to have them shipped to the site by helicopter, the supplyman said they could not put that type cargo on a supply chopper. They would have to go on one of the summer resupply ships. On that trip I had also picked up a replacement antenna for one of our vehicles. I pulled the antenna out of my parka, put it on top of the box of glasses, and said, "This goes with it." He said, "Yes sir," and made out the shipping slip for "miscellaneous electronic equipment."

Pete Baird:

My favorite Bar was always The Pastime Club in Tonopah NV. Owners Mel and Jan were great folks. Beer was always good and cold. It was a nice friendly place. I played drums there in 1964-65 with Rosie and Mary Herara.

Old Bob The Cheerful Malcontent:

There also was a Duke's Bar in Abbotsford, WI. At one time this town of about 1,400 people had about 14 bars. If you understand the people of northern Wisconsin this was normal. One of the great things about Wisconsin bars is that most will have a plate of crackers and a big bowl or wheel of Wisconsin cheddar cheese sitting on the bar for all to snack on with their favorite beer.

Donald D Simmons:

At Nemuro, Japan we were twenty seconds from Russia by jet. The location was on a bleak, windswept point of Eastern Hokkaido and we were a small band of Americans standing lonely vigil with radar lances over one of the most sensitive segments of the free world defense perimeter. Four miles away, across the frigid gray-green Nemuro Straits, laid Russian land. It was the low, flat Habomai and Shikotan islands, seized by the Soviets as spoils of WW II. The islands were shrouded in mist and mystery and farther North was the equally visible outline of Kunashiri, southernmost of the Soviet-held Kuriles. In this tense cold war proximity, perhaps exceeded only in Germany, the GI's of this desolate, frontier post had fashioned a life far removed from the bars, honky tonks and shabby dance halls surrounding most American bases.

Their unique adjustment to a semi-Arctic environment had led them to consider their duties routine. Three times they had heard big Russian coast artillery guns thunder from Kunashiri, rattling windows in their Quonset huts with practice rounds that fell harmlessly but significantly off the Japanese coast.

Unidentified boats on at least two occasions pulled up at night off nearby Hanasaki harbor and combed the area with blazing searchlights. Twice in three years a Russian fighter, within sight of the camp, had shot down big U.S. B-29 planes. Armed Russian sea and air patrols were seen regularly. 3,628 Japanese fishermen of Nemuro had been captured and interrogated for violating what the Russians feel are their fishing grounds but which the Japanese still consider their own. All but 74 had been returned. Bitterly the fishermen tell any American who cares to listen: "We want those islands back. We cannot live without them." For this reason, perhaps, the Americans and local Japanese struck up a unique camaraderie.

The airmen sent their bulldozers into Nemuro to level playgrounds for children, to grade roads and clear snow from streets that would otherwise be blocked throughout the winter. School kids cheered and waved every time a truckload of grinning GI's rattled by. In town the tough brawling Nemuro fishermen shouted lusty greetings to "on-the-town" airmen who wander into the local "Sake Alley." They often drank together through the night. Nemuro offered little to the tradition-bound American. Its muddy streets grey weathered lumber buildings, deep-chested Eskimo-like women smacked of the Yukon. Shaggy Welsh-type ponies, loaded with succulent King Crab, could be heard over the steady "chop, chop" of waves against the shore. Airmen on pass find no pseudo-American bars or dancehalls. They meet the Japanese in their own settings, usually sitting cross-legged on straw mat floors, drinking beer or sake off a low table. Capt. James A. Harris, Seattle, WA., the detachment's rugged commander, laid down a strict foundation for good relations. "It's drilled into our men that they are no better than anyone else," he says. "That's the way we operate here."

Through powerful glasses from the outpost, three towers, one possibly a radar installation, cattle, horses, and buildings could be seen on the nearest island. Japanese said soldiers in small groups were frequently observed. Powerful floodlights that could cover the straits had also been reported. Nineteen miles away the rugged mass of Kunashiri pokes through low-lying clouds. American sources had little to say about what was there, but it was no secret that both jet and conventional aircraft were spotted in that vicinity frequently. There were the constant unofficial reports of Russian agents flowing from Kunashiri and the Habomai into Hokkaido.

Airmen making this pilgrimage to Nemuro usually started back as the sun sunk into the sea, and freezing winds begin raking the stubble of long ago harvested grain fields. The Orient and the Arctic lie mixed and faded into the dusk. But a short distance ahead were the Quonsets and a warm fire that held the promise of home.

Tony Ocampo Lt/Col. Retired. How to make someone feel old:

I was stationed with 5th AF, Tokyo as a Major in 1970. Over a cup of coffee, one of my young lieutenants said to me, "Say, Major, you and I were stationed at Eglin AFB at the same time. I replied, "That cannot be, you are years younger than I am, and I was at Eglin from 1948 to 1950. He smiled and said, "That's right, I was born at the Eglin Base hospital in 1950."

Robert Chaney:

I was at Point Arena, California during the summer of 1975. Because of the energy shortage the height radars were shut down and we had much time on our hands but continued working three days on, three days off and on 12 hour shifts. I was working the late shift and was sitting on the wall over-looking the large parking lot inside the containment area, just below the height radar radome. I was sitting there with one of the sites "God squad" crew who was telling me of the advantages of living a pure and simple life. I was about to tell him that I was 22 years old, isolated on top of a damn hill in Northern CA. How pure and simple could life be? When suddenly CRACK! Lightning strikes a tree on the ridge between the main base and the housing area and it burst into flames. I think we both shouted Jesus and ran into operations and called out the fire fighter guys, which was us. By the time the on base fire truck found a driver, started its engine and found out we couldn't get close to the fire in the woods it began to rain real hard and extinguished the fire. Some of the guys from base housing also emptied fire extinguishers on the tree as I recall. Looking back on that place it was a fire trap with all the woods and old down trees left laying after that area was logged many years before we were there. I have thought many times if it was my clean living or the God squad guy. We certainly were being watched over that time.

Richard Waddell:

I was being assigned to Korea. My trip started in Springfield, Missouri in December, 1955 sometime after Christmas. My mindset at the time was to take a train. Flying was for the rich. I took a railroad coach car all the way to the west coast and slept in my seat. I did spend a lot of time in the club car and even though I was under 21. This may have had something to do with my running out of money before I got to Parks AFB California. The ride across the Sierras was memorable. Snow was very heavy at the time, and wooden sheds had been built to protect the trains from avalanches.

The scenery was awesome. When we got to Oakland, all the passengers got off the train walked down to the dock to catch a ferry across the bay to San Francisco. When I got to San Francisco, I had barely enough money for a hamburger, a cheap hotel room, and a bus ride to Parks AFB. I checked into a hotel, and since I had no money to spare, I took a walking tour of the city.

The next day, I signed in at Parks. As I went through processing, they couldn't seem to get my serial number right. Whenever I complained, the clerks told me it was nothing to worry about.

After several days, I got new orders for assignment to a unit in Japan. As we waited for the bus to the port, the NCOIC handed us our personnel records to hand-carry to the next assignment. I took a look at the one given me, and discovered that my home-of-record was now Sioux Falls, Idaho. I had never been there. The records belonged to another airman with the same name as mine. The morning report clerk had picked him up as reporting, and had me listed as being AWOL.

The First Sergeant apologized and I got to spend several more days at Parks, pulling KP, and receiving my shots again. They would not accept that I was the Richard Waddell they had given the same shots and vacinations to just days earlier. There were some real bureaucratic sadistic bastards at Parks. They did, at least, gave me the choice of going to Japan by ship, or by air. It was an easy choice because I wanted to experience what the rich folks felt.

I boarded a C-97 at Travis AFB. This was my first time to fly. We touched down at Hickam, Hawaii to refuel. It was after dark when we took off and headed for Wake Island.

The first hint of trouble was when the flight engineer came back with a flashlight to look at one of the engines. The pilot announced we had lost oil pressure in one of the engines, and were heading back to Hickham. The flight engineer again came back with his flashlight, and then the pilot announced we had lost a second engine. I was wishing I had taken the ship.

After repairs were made, we flew to Wake Island, Tachikawa, Japan and to Kimpo, Korea. We were trucked to our support base, Osan AFB, Korea. I had not been paid for six weeks but was promised that it would get straightened out.

Finally, I got to Kangnung (now spelled Gangneung), which at that time had a population of about 10,000. My first impressions were not favorable. Most of the time, at least 5 days a week, our eggs and milk were powdered.

But, on Fridays, weather permitting; a C-47 or C-119 would bring in a supply of fresh eggs and fresh milk. Saturday breakfast was always a treat. Potatoes were always powdered, and the rest of the vegetables were canned. I don't remember ever having fresh fruits or vegetables. Local crops were fertilized with human waste, so it was probably just as well.

We were entitled to two R&Rs, and I managed to work in one during my tour. We were officially entitled to seven days for each R&R, but most guys managed to stretch it to two or sometimes three weeks. They could get away with this because the transportation was so uncertain. Travel for R&R had the lowest priority, and was on a space available basis.

For our R&R, Mike Ontivaros and I flew to Fukuoka for a few days, and then to Tokyo for the rest of our R&R. Some of the guys went to Hong kong.

Kangnung consisted of several unpainted buildings, Quonset huts, and tents along dirt roads. The orderly room used crude furniture made from unpainted plywood. All of the buildings and tents were heated with small oil stoves. The squad tents looked like something out of M*A*S*H. The only grass on the site was a small enclosure around the flag in front of the orderly room. The patch was ringed with whitewashed rocks.

The toilets were also rustic. We airmen had one big outhouse with a wooden bench with about 5 holes. For the urinal, there was one long metal trough with a drainpipe at one end. The showers were fairly decent, and the water was always hot. We got mail once a week, when the supply plane came in. We had no TV or other satellite connections.

Kangnung seemed to have a large population of un-attached women. I think this was because so many men had been killed in the war or were now serving in the military elsewhere.

Some evenings women were allowed to come on to the base to visit the Airmen's Club. We would dance to records, and there was a lot of hooking-up going on. Six-bys made trips into town and back about every 30 or 45 minutes during the evening. The girl would generally take the guy back to her place in town.

On my last evening on Kangnung, I went to the Airmen's Club, and met my friend, Oley Jackson. I told him I was scheduled to leave for the States the next day, and he suggested we go into town to celebrate. "Celebrate" was a euphemism for "try to pick up chicks."

I had never met a Korean I did not like so when a group of young Koreans appeared out from a side street and got in front of us we were not afraid. They were laughing and one of them yelled for us to follow them, which we did.

They stopped in front of a non-descript building, and started up some stairs to the second floor. We entered a nice restaurant that appeared to be one of those places we had been told existed that the Koreans do not want GIs to know about. There was no sign at the entrance.

The Koreans were in their early 20's, like us. There were three men and two women. They appeared to be better well off than most of the Koreans I had seen on the streets. They spoke fairly good English, and we spent the evening with them, talking, joking, eating Japanese stir-fry and drinking sake.

I was glad to be leaving the next day. I spent an interesting year on K-18, and then retraced my route when I returned to the states. Looking back, recreation seems to have been limited. There was no swimming pool, pool table, table tennis, or bowling. I can remember no movies, but George Rodas recently told me that he showed movies during the time I was there. For recreation, we had the Airmen's Club and the beach in the summer. The Airmen's Club served 15 cent mixed drinks, and five-cent beer. That beer price may have been a Friday night special.

Dennis Radke:

When I arrived at the 731st, Sundance AFS, Wyoming as a Staff Sergeant I was assigned to the orderly room performing administrative duties. One of the things I did was to go next door and check the stuff we had stored in Supply.

The 731st was unique in that it was powered by our own teapot, a nuclear reactor. That made us a showplace for the big wigs. Visits like that would make any commander a little paranoid hoping the visiting dignataries would not find anything to level any criticism about him or the unit.

In a dark corner of supply I was directed to a huge pile of cardboard boxes containing many hundreds of pads of DD 1150 that apparently had been there just short of forever. DD 1150's are requests for equipment. It seems the teapot crew were so paranoid of running out of something, they were determined not to run out of the forms to requisition that something.

I guess I could have just started a fire and solved the problem but they were paid for with Air Force funds. In an effort to get rid of this over abundance of these forms by transferring the problem, I placed calls to Ellsworth AFB, our supply organization, and they did not want the forms. I called our bosses at Minot AFB at the Air Defense Sector headquarters and got the same answer. They had no use for them either. I used one of the requests to order a supply of parcel post sacks from the local PO and shipped the lot off to Minot over a period of days. We never heard another word about it.

Lawrence Kessler:

I was only at one site during my four years in the Air Force and that was 745th at Duncanville TX. We did not have slots in the club, rather a pin ball which we called the, "Nickel Eater". I remember it was a Bally Machine but owned by the National Amusement Company. They sponsored our Bowling Team and provided us with nice Shirts.

Unfortunately, I must admit, I was one of the machine's many customers. It could eat a roll of nickels pretty fast. Once in a while one could hit it for a few hundred games, which we could cash in with the bartender for five cents per game and he would cancel off the credits.

I worked at the club as a bartender for a couple years. I think I blew most of my little pay back into that nickel-eating beast. At a small site like ours, often times the bartender was not very busy and to pass the time, rather than watch the TV, I would partake of the nickel-eater. I might have been able to list that miserable device as an income tax dependent.

Joseph Bonta:

It was nighttime at the 708th Radar Squadron, Indian Mountain, Alaska, but then it was almost always nighttime in the winter, and I had to use the outhouse because we did not have enough water for indoor latrines. It was about 60 feet from the nearest building which was the mess hall. It must have been 40 below zero and the wind was blowing at least over 30 knots. When I was done I started to walk back but my Parka acted like a sail. I started to slide on the ice blown by the wind towards the edge of the mountain. Thinking like my life depended on what I did; I dropped to the ground onto my hands and knees and crawled to the back door of the mess hall which was the nearest door. That was the safest way to make it back. What a blessing it was to finally get the indoor plumbing back working.

We had one man who, when things got really bad, would start dancing up and down the barrack singing "ITS PARTY TIME". That meant there would be a drunk on that night. We were allowed to buy a case of beer each month. I did not drink much, so when it was party time and the guys finished their cases, I would contribute mine except for maybe 3 beers. Most guys would get so drunk they could not pull their shifts at wherever they worked. The few who did not drink or had not got drunk would cover for those who did. I pulled scope dope, cook, and driver, whatever had to be done so the site would remain on line.

Living conditions were hard at Utopia. We had flooded barracks from broken pipes from the minus forty-degree weather. There were times when the weather kept us from going to bottom camp, let alone Fairbanks. The only

227

good thing was the food. But if we did not get resupplied we had to resort to "C" rations. There was one incident where the outhouse blew over and no one noticed until someone realized we were missing a man. A search was started and saw that not only a man was missing, so was the outhouse. When we looked for the outhouse and found it we also found our missing resident. We solved that problem by resetting the facility and tying in down with steel cables.

The biggest gripe I had about Utopia was the lack of an indoor latrine. That outhouse was not too bad during the summer and late spring and early fall, but it was not too nice when the temperatures were 10 to 60 below zero. The way it was set up when you had to use the outhouse, your butt was really exposed to the outside temperature. Inside the outhouse was a potbelly stove that kept the temperature inside above freezing. When it came time to clean your butt it was normal practice to bend over with your butt to the potbelly stove to thaw things up so you could use the toilet paper. If other guy's were in the outhouse you just did your thing and had a conversation with whoever was there at that time. It was so common that I nor anyone else thought it was strange.

Jim Arbanas:

I had a similar experience while serving with the 911th at Fort Concord, Vermont I had taken a bag of trash outside to our secured dumpster. Wind never stopped blowing up on our topside operations area and I remembered readings that night of 40 to 60 mph plus winds and temperature reading 30 below zero. Like you, my arctic survival parka acted more like the sail you mentioned with me being blown off my feet and towards the edge of the mountain. My "pucker power" shot way up as I was shoved across a hundred feet of packed ice nearly a foot thick. The wind subsided just enough, and long enough for me to craw back to the entrance of the operations building and fumble with our security coded entrance before the next big blow started.

B. J. Roke:

The favorite brand of cigarettes for a lot of enlisted men was O.P.'s (other peoples) and a lot of guys who smoked got kind of tired of supplying to folks who never returned the favor.

While cleaning up our barracks at Montauk AFS for an inspection, we found a duffle bag with letter in it that was dated about 1958, this was about 1963. Also, in the duffle bag was a full carton of Pall Malls and the guys who smoked took a couple of packets for the folks who smoked O.P.'s. I didn't smoke, but from what I was told these things tasted terrible and burned like a fuse. For a while after that cigarette mooching went way down.

Bob Wilder:

While stationed at Red Cliff, New Foundland with 108th AC&W I had a coffee pot in my room. Seem like every time I brewed a pot everyone came looking for a cup. I got tired of that so I got on the MARS station and one of my buddies, Boyse Medera, contacted his mother down in the bayou of Louisiana and had her send up a can of chicory. When it is brewed it smells great but it is has the strongest flavor imaginable. When the bummers came around for a cup and tasted chicory, most of them could only drink a small part of a cup and quit. They also quit bumming. I got to enjoy it. If you ever get the opportunity to taste chicory, do so but make certain there is some drinkable water nearby.

Ray Buda:

I had an experience back in 1966 as part of the General Electric engineering team that replaced one type of radar with another at Wallace Air Station in the Philippines.

I was selected to ride shotgun with the commercial truck bringing the equipment from Clark Air Base to the site at Wallace. Everything went smooth until we took the antenna sections to the site. From the earlier trips we knew it would mean leaving Clark in the morning, getting to Wallace, unloading the truck and getting back to Clark the same day. That was of course if Mr. Murphy cooperated. Mr. Murphy has a law that if anything can go wrong it will and at the worst possible time.

On the day that the antenna sections were to be transported I was told that at one of the bridges due to weight restrictions most of the sections would have to be down loaded, dragged across the bridge and reloaded. I thought that would be no problem, I was wrong. When the second section of the antenna was loaded on a truck the banding equipment broke and they wrapped banding material around the antenna crate and secured it to the trailer. After that, three more sections were loaded and the same thing done. That meant at the bridge we had to un-band the four crates, unload them, drag them across the bridge and reload and re-band them to the trailer. Not a simple task, but we did it.

Finally we got to the site 24 hours after we left Clark. When we got to Wallace they had a sign over the base entrance that was lower than the loaded cargo was. The commander strongly suggested that we flatten the tires to get the load under the sign, drive under it then re-inflate the trailer tires. I had a better suggestion which as a civilian I could make. I had a man from civil engineers cut one side of the sign, lifted it, and the trucks drove through. Much simpler to cut and re-weld the sign than it would have been to let the air out of eight tires, drive 80 feet on the rims and hopefully not lose a tire and then re-inflate.

When we got to the radar tower and unloaded the antenna sections my boss told me that I should be ready for work. I told him to kiss my grits as I had been up for over 24 hours and I was going to take a rest. I did get my rest after having a couple beers to mellow me down. Mr. Murphy let us alone the next day when the tower was put together.

Jim Sanders tells of this story:

A US Air Force C-130 was scheduled to leave Thule Air Base, Greenland, at midnight. During the pilot's preflight check, he discovers that the latrine holding tank is still full from the last flight. So a message is sent to the base and an airman who was off duty is called out to take care of it.

The young man finally gets to the air base and makes his way to the aircraft only to find that the latrine pumper truck has been left outdoors and is frozen solid, so he must find another one in the hangar, which takes even more time. He returns to the aircraft and is less than enthusiastic about what he has to do. Nevertheless, he goes about the pumping job deliberately and carefully (and slowly) so as not to risk criticism later.

As he's leaving the plane, the pilot stops him and says, 'Son, your attitude and performance has caused this flight to be late and I'm going to personally see to it that you are not just reprimanded but punished.'

Shivering in the cold, his task finished, he takes a deep breath, stands tall and says, 'Sir, with all due respect, I'm not your son; I'm an Airman in the United States Air Force. I've been in Thule, Greenland, for 11 months without any leave, and reindeers' asses are beginning to look pretty good to me. I have one stripe; its 2:30 in the morning, the temperature is 40 degrees below zero, and my job here is to pump crap out of an aircraft. Now, just exactly what form of punishment did you have in mind? Are you going to transfer me to a radar site? Well Sir, right now, that don't look too bad.'

REAL WORLD AIR DEFENSE OPERATIONS

There were many real-world "teeth clenching" times experienced by those manning our air defense radar sites. Professionalism and strict dedication to duty enabled our USAF Air Defense forces to operate as planned and protect the world from nuclear war. The following stories document the reason the entire air defense system was built in the 50's and expanded & improved in the 60's – to counter the operational threat.

An important addition to U.S. air defense system was the EC-121 four engine long-range radar surveillance aircraft. In addition to providing extended radar surveillance off the East and West coasts of the U.S. the AC&W aircraft also flew off of Iceland controlling USAF fighter intercepts of TU-95 Bears. The EC-121's carried a crew of 11-25 surveillance and intercept control personnel in addition to a six-man flight crew.

An Insider Viewpoint- Soviet TU-95 intercepts by Stan Fields:

I spent many TDYs in Iceland, from 1971 to 1979, while assigned to the 552nd AEW&C Wing from McClellan AFB, California and Det 1, 20 ADS, Homestead AFB, Florida, flying the EC-121s, The "Connies". Our NATO Mission in Iceland was to detect and intercept TU-95s and other Soviet Bloc military aircraft. We did this very well with help from the Brits & their Lancaster AEW&C Platforms.

Two experiences with TU-95 intercepts are as follows:

We were scrambled due to intelligence suggesting that TU-95s were about to penetrate the ADIZ, we chased them all over the sky, controlled successful intercept activity as did the Brits numerous times. The end result being that right after we landed at Keflavik, Iceland (KEF), the TU-95 also recovered at KEF for maintenance and refueling. Then both our crew and the TU-95's crew spent the evening together at the Chief's Club at KEF. They departed early the next morning.

We had the same scenario as above but there was no TU-95 recovery at KEF. Again, after many successful intercepts the TU-95 flew with us, side-by-side for a half-hour or so with the fighters from the 57th FIS trying not to fall out of the sky as they escorted us until the Bear decided to leave with a Courtesy "Wing Wobble" and a hand salute from the cockpit crew.

An Insider Viewpoint- pictures of Intercepted TU-95's as told by Art Mussman:

I was the Operations Officer at radar site at Rockville, Iceland in 1967. The 57th FIS were flying F-102s at that time. They were doing many intercepts of Bears and taking pictures for intelligence purposes, but the quality was mediocre, primary due to the pictures being out of focus. The cameras were hand-held, and many pilots knew little about photography. The squadron started an upgrade program featuring photography lessons for the pilots and new Nikon Single Lens Reflex cameras. The quality improved except for focus. It was finally determined that it wasn't focus, but the eight counter-rotating props on the Bear were seldom in sync, causing the whole airplane to vibrate faster than the lens speed. The solution was high speed film and a faster setting on the camera.

The Soviet heroes of the Cold War were the Bear crewmen, who endured 20 plus hours of this constant shaking on each mission. I imagine it was hard to keep fillings in their teeth!

An Insider Viewpoint- The Day We Ran Out of Fighters as told by Les Crine:

It was the middle of May, 1969 and I was working the Weapons Assignment Technician (WAT) position at Campion AFS, Alaska. I had just finished issuing Active Air Defense scramble orders for the 5 and 15 minute alert birds from Galena AFB. We had multiple "pinball" contacts from our Distant Early Warning (DEW) Line sites and multiple "unknown" radar contacts from our Ground Control Intercept (GCI) sites at Tin City and Cape Liz. According to our electronic plotting board, it looked like every GCI site and every DEW Line site in our Area of Responsibility (AOR) was being targeted. It

234

didn't take a genius to realize that we didn't have anywhere's near enough fighters to handle this situation.

The next thing we knew King Salmon had scrambled all its birds as had Eielson. Some where's in the mess I think I heard that Elmendorf was deploying more fighters to King Salmon and Eielson and that there was talk about bring the F-106s standing alert at McCord AFB, WA into the fight. Next to my position I had a piece of equipment that kept track of the entire fighter force on alert or Operational Ready (OR) within the Alaskan Air Command. As each set of fighters were scrambled and became airborne, the respective WAT would enter the change in status for his base or bases into the system. One thing was for sure, the number of airborne fighters was increasing and the number of aircraft on alert was fast approaching zero. A look at our plotting board confirmed that. The pucker factor was definitely on the rise this day.

When the number of fighters in the Airborne and OR status columns reached zero things really got interesting. The "unknowns" our GCI sites were reporting showed no signs of breaking off and neither did all the DEW-Line "pinballs" (position of unknown aircraft reported by teletype from DEW radars). The Alaskan NORAD Region had scrambled every F-102 that was capable of flying and fighting and we still had "unknowns" and "pinball's' that were carrying a TAC ACTION Status of "AD" (Action Deferred) or "NFSL," (No Fighter Suitable Location). This was defiantly not a good position to be in.

Slowly but surely, things started to change. One by one, the "unknown" and "pinball's" started to break off their runs against our sites and change their headings away from the coast. It wasn't long after that and we started to recover our fighters. Within those couple of hours it sure looked like a different world. And when it was all said and done, everyone agreed that we had a "three sweat-ring day." I heard later that the Intel folks at Elmendorf and NORAD believed that the Russian Air Force was in the process of re-deploying from their winter staging bases back to their home bases and they decided to have some fun at our expense on their way home. Some Fun!

An Insider Viewpoint- Alaskan Intercept Operations as told by Shaun Finn:

In Nov 61, I was transferred to the 748th AC&W Squadron, Kotzebue, Alaska. This was the most northern GCI Site in the Alaskan Air Command. Our MDC was at Galena. During my year there we had numerous "over-flights" and "bubble checks" by Soviet aircraft. On only one occasion were we successful in intercepting the aircraft and chasing it back over the ADIZ. I was an ICT during my year there. The F-102A interceptor aircraft was a poor choice at best for keeping tabs on Soviet penetrations. AAC required external fuel tanks for the F-102's that limited their speed; available launch bases, and limited their range. Silent Sam Scrambles were necessary as Soviet long-range radar could pick up our efforts (mostly US Navy) to intercept them. Low Altitude scrambles consumed too much fuel, and prevented a chase if we needed it. The standard F-102 carried a mix of IR and radar guided weapons. We transitioned to the GAR11 that was a nuclear armed guided missile that replaced the old MB1 Genie.

The Cuban Missile crisis was a time of intensive air defense operations. When the crisis was over there was no change in the need for vigilant air surveillance by our radar sites. This was the standard maintained by Air Defense Forces whose 24x7 mission throughout the Cold War was to detect, intercept, identify, and destroy when necessary.

An Insider Viewpoint- Soviet Aircraft Detected, Intercepted, and Fortunately Identified as told By Bernie Roke:

I was a heavy radar tech at the 773rd Radar Squadron, Montauk AFS, NY from Oct. 1961 to Sept 1964. Shortly after the end of the Cuban Missile Crisis in early December, 1962 one of the radar operators told us that they had been instructed by the SAGE Direction Center at McGuire AFB to look for a target at the extreme end of our range in the southeast sector - possible Soviet aircraft. My first thought that someone was trying to keep the pot stirring after we had gone back to Defense Condition 5 (Normal Readiness).

Shortly thereafter a target appeared in the area that McGuire had called out and a couple of F-101B's from Suffolk County AFB went out to have a look. After a while, all we got back was "Positive ID - Soviet Aircraft". That got me a little spooked. A few days later, I found out the target was a Tu-114 Aeroflot airliner en route to Havana, Cuba. The Tu-114 used the same engines, wings, and empennage as the Tu-95 and we painted it every sweep at maximum range.

Prior to the Cuban Missile Crises, Aeroflot Tu-114's (the civilian version of the military TU-95 Bear) flying to Havana would regularly stop for fuel at Goose Bay, Labrador. They were able to stay well to the east on the leg from Goose Bay to Havana and so we did not detect these flights. When the crisis began, the USA asked the

Canadian government to cancel the landing rights for Aeroflot, which they did. So after the crisis ended, the landing rights were not initially reinstated and the Tu-114's would fly to Murmansk, refuel there and fly non-stop to Havana. The shortest route would bring them inside our ADIZ but outside US territory and this resulted in intercepts to check them out. After that first event, the radar operator logs would indicate a Tu-114 flight on an regular basis. There were unconfirmed stories that these flights made for some very hairy over TU-114 gross weight take-offs at Murmansk and fuel critical landings in Havana.

Among the weapons available for air defense was the CIM-10B BOMARC nuclear armed surface-to-air missile. The BOMARC missile was powered by twin ram-jet engines. Upon launch, it would climb to 70,000 ft, and was guided to the target at twice the speed of sound by an Intercept Director in the SAGE Direction Center. Engagements could take place more than 440 miles from the launch point, generally a site on either the East or West coast. Once in the vicinity of it's target, the BOMARC would dive on its target. Terminal guidance was provided by an internal radar system and, with it's nuclear warhead, it could be used with devastating effectiveness against attacking bombers.

An Insider Viewpoint- An Inadvertent BOMARC Erection as told by Bill Peterson:

I was a Weapons Director for BUIC II (Cape Charles) and a Senior Director for BUIC III (Ft Fisher) and we qualified for Senior Missile Launch Officer badges. We went to USAF BOMARC A and BOMARC B school for formal training. We also had control of the Army NIKE sites in the ADA rings around Washington, Baltimore and

Norfolk, plus the Navy weapons that chop to CINC
NORAD should we ever reach DEFCON TWO/ONE or
Air Defense Emergency. These weapons were all
nuclear warheads (small) and were primarily designed
for use against Soviet TU-95 BEAR long-range bombers.

One day the BOMARC unit near McGuire AFB was
having an evaluation and for some reason the missile
went into launch sequence. You have to visualize this:
the concrete building splits open like a V, the missile gets
upright, and if commanded correctly it takes off toward its
target. It got all the way to the vertical stage and was
shut down. The unit reported it as "an inadvertent
erection".

Radar surveillance of the airspace being defended had
many other benefits. Friendly aircraft could be in trouble
and the controllers at radar sites could offer a helping
voice and directions to a safe landing.

**An Insider Viewpoint- Saving a U-2 as told by Art
Leighton:**

While in Alaska, in 62-63, I noticed a U-2 varying from
his route, and heading towards Siberia, I made calls in
the blind to get his attention, and finally got him into the
code change routine for Yes and No on his IFF, and
determined he also had lost all his navigational gear, and
radio transmitter, and he was low on fuel, and
disoriented, so I got him turned around in the direction of
Galena on the West coast of Alaska. The pilot started a
spiraling decent to that field and at about 35,000 ft. he
flamed out, and glided to the runway for a no power
landing.

An Insider Viewpoint- KC-97 In Trouble as told by William Shaw:

From 62-64 I was an AST at the 26th Air Division, Hancock Field, NY. My duty position was just below the Battle Staff. We had a KC-97 one swing shift on "flight-follow" through our northeastern US area. I believe his flight originated out of Plattsburg AFB, NY located within the physical boundaries of our Div., but I'm not sure on that. He was doing night refueling missions.

We paid special attention to him when he started squawking an emergency signal. As he proceeded to exit our Div. boundaries heading North by Northeast, we learned his emergency was a possible fire on board as indicated by one of their on-board alarms that monitor their fuel tanks/and lines and such. The pilot ordered the crew to bailout then set the ship on "auto-pilot. The aircraft never did catch fire. It never did crash until it ran out of fuel way North. Sad part was the parachute of one KC-97 aircrew member failed to open and he perished.

That was nearly 50 years ago and that is how I remember the incident. Throughout this incident there was heightened concern and interest. I remember very well the Officers on our crew huddled around our surveillance consoles

An Insider Viewpoint- My First Intercept Mission as told by Les Crine:

Radar intercept control against enemy aircraft is very intense, especially if it is your first time.

The exhilaration that comes with doing a good job and being recognized for it makes up for all the effort it took to get to this point in your Air Force career.

Shortly after arriving at Ft. Yukon, Alaska, my Crew Chief in Surveillance discovered I was color blind. Since tracks were manually plotted on a vertical Plexiglas plotting board, color coded based on identity, this meant I could no longer work surveillance. As a result, I was banned to the Operations Office orderly room to perform clerk duties. Since I was the junior man, I usually drew phone duty during lunch hour.

It was on one of my lunch hour phone watches that my Air Force career took a big change. As usual, I was alone in the Ops Office during the lunch hour, when the Senior Director (SD) called in on the intercom system and ordered all personnel to report to the Ops Room on the double. He had two "Silent" scrambles and he needed bodies right now. When I told him nobody was there but me, he ordered me to get my ass in here on the double, NOW!.

As soon as I entered the Ops room, the SD told me to report to Capt Carroll on the top dais and work as his Intercept Control Tech (ICT). Christ, I couldn't believe I was going to work a live intercept mission. Not only live, but against Russian Badgers. I didn't know squat about being an ICT. I reported to Capt Carroll, he told me to put on a head set, break out the correct NORAD forms to record the mission, write down everything that happened, everything, and to take care of all his comm needs.

By the time the crew members reported in from the mess hall, Capt Carroll had two flights of 2 F-102s identified, under his control and was fully engaged in running two "hot' intercepts against two flights of Badgers that weren't even on his radar screen yet. He was running the intercepts based on the "Pinball" reports off the teletypes, received from DEW sites, which I kept feeding to him. When one of the Crew's qualified ICTs stepped in to take my place, Capt Carroll told him, "no, leave Crine alone, he's doing just fine." Man, I felt 10 feet tall.

When the mission was over and the fighter's were returning to base, Capt Carroll reviewed my logs, told me I did a good job and asked me if I wanted to train to become an Intercept Control Technician (ICT). Without hesitation I told him yes. He said he would talk to the Commander and see what he could do. The next morning when I reported to work, the Ops Officer told me that I was back in Ops and that I was going to train to become an ICT with Capt Carroll and work days. That was without a doubt, one of the best days of my entire Air Force career.

After that I stayed in the Weapons or Control Sections for the next several years, but I'll never forget my first mission. Some guys, enlisted or officer, would go an entire career of 20 years or more without ever experiencing the feeling one gets when running a live mission against a Bear or a Badger or any bad guy for that matter. I was lucky; I did it on my very first mission.

PRANKS AND PRACTICAL JOKES

Pranks and Practical jokes were played on many as a way to interrupt the boredom of life on the sites. Rank was seldom considered when the perpetrators offered up their wares. In fact the higher up the chain of command a prankster could get, and not get caught, the better the prank. These are just some:

Chief Master Sergeant Wayne Fitzgerald:

My first night on the Goose Bay site in Labrador I met these airmen who unbeknownst to me were the two pranksters on the site. They came up to me right after I had been crowned King of the Club. One of them said. Hey Sarge, we will flip you for a quarter; I said OK and started digging in my pocket for a quarter and they both grabbed me and flipped me on my hinny. Of course everyone got a big laugh out of that, but from then on I kept an eye on the rascals. Another trick they liked to do was catch someone passed out in the club and roll them out across the hall and into the area of the latrine that housed the washers and dryers and prop them up in front of one of them old Westinghouse front loading machines that was running and just leave them there like he was watching a TV.

It was rumored they had shoved the first sergeant down the stairs toward the enlisted wing from the club and it accidently broke his leg. I met him at Goose Bay when I was going in and he was hobbling around on crutches and he warned me about that pair. One of the two and I actually became friends and he accompanied me several times to the fishing camp on Sandy Point.

Another trick they played was to sneak into someone's rooms when they were asleep. They would pick them up, cot and all and take them outside. They would then either dumping them off the cot or just leave them outside. They would also take them from their room and put them in someone else's room. They tried that on me not realizing I was a very light sleeper. I waited until they had my cot tilted to get through the door and I smacked one of them between the eyes with a round granite core sample, about the size of a roll of quarters that I had in my hand. He was laid out cold as a cucumber. Never again did either one of them try their practical jokes on me. It actually slowed down the antics of the two with others.

Jim Sanders offered this:

When we pull a good practical joke it is of little value unless you can take credit for it.

It must have been some time in 1959 at Birkenfeld Germany. I was in the mess hall eating supper and my next stop was usually the NCO club for a drink then back to my room to write some letters. There was this rather large bowl of peanut butter as part of the food offering. I had an instant idea. In the dining room at the club there was a childs high chair with a potty seat built into the high chair and the catch bowl was clear either glass or plastic. So I took a healthy amount of peanut butter wrapped in a napkin and off to the club I went. I had rolled it to look like a BM and went into the club. It was early so no one was in the dining room. I went in and ordered a cup of coffee from the only staff member in the room. Quickly, while her back was turned I deposited my rolled up peanut butter object into the transparent catch basin of the high chair. When my coffee came I took it to the bar and ordered a beer. It was maybe an hour later all hell broke loose in the dining room. People where saying things like "anyone who would allow their child to do that in the dining room," and things like that." The waitress, who carried the chair, holding it at arms length in front of her, had this real disgusted expression on her face. Bottom line is I could not take credit for the joke – someone might have killed me.

Mel Hoke:

On practical jokes, a favorite played on a newly assigned person at Miles City AFS, Montana involved rattlesnakes, which were numerous on the site. One of their most common "hangouts" was in the cable trenches from the tower to the operations center. In fact, maintenance checks in these trenches were such a high risk that they should have qualified for hazardous duty pay. We had killed several rattlers and skinned them to be used as hat bands but we did have another use for some. We would coil them with the head propped up and put them in the path of a new guy to enjoy their startled reaction. Several times we saw a new guy actually leave earth defying gravity. His feet would be moving but he would only be going straight up then straight down, straight down to the snake. Then he would be making dust.

When I went to Moriarty AFS, New Mexico a practical joke in the mess hall was to offer a new person what looked like a regular sweet and sour "mini" pickle but that was actually a Mexican hot pepper. I fell for that one and everyone had a good laugh seeing me swallow that "hook line and sinker" and go into a coughing fit as I tried to recover. The bottom line is that practical jokes are fun for the perpetrators but not for the victims.

Michael Murphy:

When I was at Sparravohn, Alaska there were stories that the Operations building was haunted by the ghost of a man who was killed there. During the midnight shift, there were just three lower ranking airmen and one of us had to make a fire check every hour There was always a battle to get someone to do it because of the site ghost which we believed was real. One night I was on scope, another was taking a nap up in the weapons level that left one guy to do the fire check. Shortly, he came back and was really nervous. I was still on scope and he asked me where the third man was. I just pointed towards the dais saying over there, at that time the napper looked up and all you could see was his head. The dim lights caused him to look pale. The fire check dude only saw was a ghostly head floating in mid air and whoosh off he went. We had a hell of a time getting him back in ops.

A stranger thing happened the next night when we were telling and showing him what he seen, how it happened and that there is no such things as ghosts. The back door to the outside opened then closed with no one coming in or out. The three of us looked at the door thinking if we actually saw what we thought we saw. Five minutes later it opened and closed again. Now this is in winter when that door is a pain to open because of the build up of yellow snow. There is so much build up the door has to be kicked or forced open and closed but this time it just opened and closed like it was well oiled. We went over to check the door and tried to open it, it was its normal stuck self.

For the rest of the night the three of us did not say too much to each other and were more happy than usual to see the day crew come in. We never made any more ghost jokes

Frank Marsh:

This is more of a confession than a practical joke. I remember seeing the movie "Stalag 17" and thinking guys could not really be that crazy or that funny. Then I went to Detachment four of the 608th AC & W Squadron. It was located on a Korean mountaintop. I really do not think some of our actions were funny at all now but it was a different time and place. One of our radar maintenance guys would just not get up and out of his sleeping bag. It was after all cold and his sleeping bag was warm. Now we can all sympathize with him because we all have done the same thing, but not every morning.

It was late one morning and the rest of us were up and he was still nice and warm. Three of us picked him up and threw him and his sleeping bag over the mountainside. With the angle of the mountain and the snow covering the ground he slid quite a ways. Then he had to get out and get back up the mountain barefoot. It may not have been funny to him then but to us it was hilarious.

Bernie Morris:

We all know some folks who sleep so hard you could drive a tank around in their room and they would not wake up. We had one of these. We had tried the warm water trick, putting his hand in the warm water to make him wet the bed but that did not work. So the practical joke team went into conference for our next move. We got a condum filled it with water so full it was quite hard to get it from the latrine to his bed without breaking. We broke one about two feet from his bed. He did not wake up and we had to go outside to laugh but like the good jokers we were we went back into the latrine, got another condom filled with water and returned to the scene of 'the crime.' He had his blankets tucked in which made it hard to get his blanket back far enough to deposit the condom in his bed with him but we did. All the time we are about to split our sides laughing because he still was not waking up. We got it under the covers and on his stomach, still nothing. We had moved away to see what would happen and he was sleeping right through the ordeal. One of my fellow jokers just punched a hole in the condom and it put about a gallon of water on him. That woke him up.

Jim Sanders, made another offering:

It does not pay to be a heavy sleeper and go to sleep with a bunch like us around.

Here is the plan - we knew this guy slept like a log. We were going to wait till he got to sleep them carry him bunk and all outside the barracks and see how long it took him to realize that he was outside. To pull this off we got four guys to slowly pick up the bed.

Check he is still asleep. Down to the end of the barracks - check still asleep. Now one guy starts laughing we have to set him down go outside and compose ourselves - which we did then back into the barracks and pick him and his bunk up and started toward the door. The bed was too big for the door so we slanted it and were half way out the door when the bunk rolled over and he fell out of the bed and onto one side of the door opening. That was what it took to wake him up. Sure, he got mad and promised pay back but I left before he got me back. I don't know about the other guys.

These jokes from others keep reminding me of other things we got into on my shift. We would get a new man assigned and tell them we were taking up a collection of money for "The wife of the Unknown Soldier". We all know GI's have a big heart and some guys would make the move to donate. When they did we chided them that if the Unknown Soldier is unknown how could we know who his wife was. Then we all would have a good laugh.

Tom Bradley:

Texas Tower #2 was located approximately 110 miles east of Provincetown, MA. Tower #3 was located off of New York and tower #4 was located off the coast of New Jersey. Tower #4 went down on January 15, 1961 with a loss of 14 military and 14 contractor personnel. In 1962 select members of Texas Tower #2 & #3 were sent to Submarine escape school at the submarine base in Groten, CT. A capsule was provided to each tower and seven men nicknamed the "Suicide Seven" stayed aboard the towers to ride out any severe storms. Once I was part of the Suicide Seven and we were ordered into the capsule and dogged the hatch. We rode out the worst part of the storm sitting in the capsule. It was a good thing we were all really, really scared as there was not a latrine inside of the capsule.

Roger Jackson:

Several of us were in our cups feeling sorry for ourselves wanting to know why the hell we were sitting in the middle of nowhere. It was quiet in the club and there was a phone. The more we talked about it the more certain we became that only one man could answer the question. So using the autovon network we called the white house. President Nixon was busy but we did get to talk to Vice President Spiro Agnew who sounded as drunk as we were. We got an answer. We were there protecting the United States.

Ray Looker gives us this final story:

Who says Santa Claus doesn't exist.

It was back in 1955 when a Colorado Springs newspaper printed an advertisement for Sears and Roebuck. In the ad they listed a telephone number telling the children if they wanted to talk to Santa to call the number. Fortunately or unfortunately, depending upon your position, the telephone number was one digit off. The listed number was for the Continental Air Defense Command, the forerunner of ADC. On duty was Colonel Harry W. Shoup who answered the phone and talked to a little boy.

The boy thought he was talking to Santa himself and Shoup thought it must be some kind of code. But Colonel Shoup was a nut about Christmas and soon realized the little boy thought he was talking to Santa. Not wanting to break the kid's heart, Shoup responded with a hearty "Ho, Ho, Ho."

For the rest of the night call after call came in and they were answered with an appropriate "Ho, Ho, Ho." Colonel Shoup had to stop for a while to get the problem resolved and had the calls redirected to another officer to take over for him.

The problem taken care of the men sat around and talked coming up with the idea of having Santa tracked on their radar system. If Santa did exist and he had a flying sleigh and reindeer, he would have to be able to be tracked by radar and who better to track than they, the Continental Air Defense Command. The idea was approved and the fun began. Onto the 60 by 80 foot glass map of North America, they plotted the return from Santa's sleigh starting on the east coast of North America and travelling to the west coast. Eventually they plotted the entire world. With the age of computers it was possible to track the fat man starting at the international dateline on Christmas evening. He has visited the Seven

Wonders of the World but has never filed a flight plan. Only he knows where he will go.

In 2008, the Santa tracking center answered 94,000 calls and responded to 10,000 e-mails.

Approximately 10.6 million visitors went to the web site which can be viewed in numerous different languages.

At Duncanville Texas, at the 745th Radar Squadron, Phil Harrison who retired in 1976 and a few others who could draw fairly well created an etching of Santa and sleigh. They then overlaid it onto a radar screen and with each sweep the image was moved closer and closer. The local television station put the image on television for several years then a more high-tech system was designed and used.

Today, more than a half a century later many of the old war horses of the Air Defense Command turn on the television on Christmas Eve and watch the progress of Santa Claus as he streaks across the sky. We tell our great grandchildren how radar can track Santa. Then we smile and know that we were once part of the organization which, with our help, started the whole thing.

So the next time you are asked if Santa exists, you might answer "If he doesn't, who is being tracked by radar every Christmas?"

However Christmas time like other holidays for the airman at the radar sites was a lonely time. Generally small private pity parties are held which usually started and ended at the club with small groups of men who became very close friends.

COMMENT

Jack Miller, Author:

I was part of the Air Defense Command (ADC) and served at two radar sites and two ADC fighter interceptor bases. My job as a Security Policeman was access control, to make certain that only authorized personnel had access to our personnel, equipment and installations. We were called many names, Sky Cops, Air Cops, Air Police, Apes and some other names not so nice. At some assignment we called ourselves Ramp Rats because one of our responsibilities was guarding the flight line where the aircraft are parked. In all seasons and in all types of weather, be it 40 degrees below or 100 degrees above zero, we were there checking identifications. Our enemy was saboteurs from the USSR. But our immediate enemy was the wind, which we called the Hawk. If the hawk was out it would be a miserable shift. Other Security Policemen were guarding a second and in some case even a third ring of security for specific areas such as weapons storage areas to prevent sabotage. In Air Force security there were only a few indoor posts. One such indoor post was at ADC radar sites guarding either the entryway to the site or entryway to the operations area where the radar equipment and personnel were.

These men were heroes in their own right, but not all heroes are awarded medals. Many heroes quietly go about their business, being there, being alert, and keeping the peace. These have been some of the stories of the unsung heroes of an unsung war, the Cold War.

EPILOG

The airmen who served at AC&W radar sites during and after the Cold War, look back on their experiences and make the claim that they survived. Most feel they are the better for the experience. Isolated locations, extreme bad weather, family separations, 24/7 year round operation, missed holidays, training exercises, and the lack of modern diversions were met and overcome. There were severe hardships on the men and their families. On the other hand, there was nothing like the satisfaction of being a member of a team that accomplished its' mission day-to-day, 365 days a year, over the course of years. Many a young airmen grew up to become men on an AC&W/radar site and will say so with pride. Today, internet groups such as radomes.org and the YAHOO Groups like the Air Force Radar Sites Veterans, and All Services Radar Vets have large, active memberships and post messages about past experiences, and provide documents and photos of earlier times at a radar site.

Today the Air Defense Command is a thing of the past. It has gone into history. Gone are the ADC Fighter Interceptor Bases, their fighter aircraft and the crews. Gone are the radar sites and the men and women who served. Some of these sites were converted to other uses such as juvenile detention centers while others were just razed to the ground. Some were saved and are in use by the Federal Aviation Agency manned by civilians.

There are former sites where there are still buildings and others which have been returned to their natural form. On such is the former site approximately 12 miles west of Winslow Arizona.

That site had been leased to the USAF by the Navajo Indian Tribe. On top of a mesa was the 904th Radar Squadron. It was a one level site consisting to two barracks, supply building, office, operations building, two towers, power plant, pump house, water tank, 21 single family homes, a community center and a gate house. Also on top of the mesa was a small arms weapons range and a SAC Radar Bomb Scoring unit. In 1963, the site closed.

Initially the site was used by another government agency then was given to the State of Arizona to be used as a youth camp. In approximately 1970 it was returned to the Navajo tribe but the tribe refused to accept it until it was returned to its original form. The government destroyed all the buildings and the buried the stuff that could not be removed. On the way off the mesa the last vehicle tore up the access road. It had been returned to its natural state except for the ghost of the young civilian girl who died in 1962 when she fell off her horse. The ghost of the electrician who was working on some high tension lines and stepped back on the platform right into other high tension lines and was killed in 1963 is also still there. Gone is the blood from the Security Policeman who accidently shot himself in the leg while in 1963, he was playing with a revolver. Gone are the memories, good and bad, stemming from parties and dances at the NCO club.

However, while it, and all of the other sites were up and running, manned by the members of the USAF, they had done their job. During that time of the cold war, we were not attacked. The men and women of the Air Defense Command assigned to perform their duties at radar sites, remote, isolated, near or far from civilization, in the United States or overseas, endured hardships and stood their posts while the rest of America slept.

Another former radar site, home to the 664th Radar Squadron is located near Bellefontaine, Ohio. This is now the home of the Air Force Radar Museum. As of this writing it is still under construction much like the radar sites of old. It seemed they were also a work in progress right up to the time they were closed. If you are ever in the area of Bellefontaine Ohio, please visit. Say hello to the veterans who will be working the site and remember to thank them. Please do not bother the ghosts which may or may not roam the halls.

The Peacekeepers

APPENDIX I

DEFENSE READINESS CONDITIONS:

Each airman had to know what defense condition they were at. This was used to determine what security precautions they had to take if certain events occurred. These are the Defcon levels and what they meant. Included are the Defense Condition Codes used during practice or exercise.

Actual: DEFCON 5 - Exercise: FADE OUT
Condition: Normal Readiness.

Actual: DEFCON 4 - Exercise: DOUBLE TAKE
Condition: Increased intelligent watch and strengthened security measures.

Actual: DEFCON 3 - Exercise: ROUND HOUSE
Condition: Increase in force readiness above normal.

Actual: DEFCON 2 - Exercise: FAST PACE
Condition: Further increase in force readiness but below maximum.

Actual: DEFCON 1 - Exercise: COCKED PISTOL
Condition: Maximum force readiness.

The Peacekeepers

APPENDIX II

UNIFORMS

In the 1950's and 60's enlisted recruits to the USAF were
issued clothing from supply. Until they received their uniforms
they were called rainbows because of all the different colored
civilian clothing they wore. This is a list of the clothing issue
each man received and the cost.

Item	Qty	Cost
Bag, Duffel	1 @	$4.20
Drawers, cotton, white	6 @	.67
Undershirt, cotton	6 @	.74
Socks, wool, black	4 @	.64
Socks, cotton, black	4 @	.35
Gloves, inserts, wool	1 @	1.19
Gloves, shells, leather	1 @	4.46
Insignia, cap, EP	1 @	.32
Insignia, Collar, U.S.	1 @	.29
Insignia, Grade, Amn 4"	8 @	.23
Insignia, grade, Amn 3"	3 @	.15
Insignia Tape, USAF	5 @	.07
Necktie, wool, blue	2 @	.69
Belt	2 @	.32
Buckle	1 @	.25
Towel, Bath	2 @	.58
Shoes, dress, black	1 @	9.83
Shoes, service, black	1 @	12.20
Boot, combat	1 @	12.50
Shirt, poly, blue	3 @	3.15
Shirt, poly, tan	3 @	2.35
Shirt, utility	4 @	2.92
Trousers, wool, poly, serge	1 @	7.50
Trousers, wool, poly, Tropical	1 @	6.26
Trousers, poly, tan	3 @	5.89

Trousers, utility	4 @	3.76
Coat, wool, poly, serge	1 @	33.30
Coat, wool, poly, tropical	1 @	6.26
Raincoat	1 @	12.30
Cap, service	1 @	6.25
Cap, garrison	1 @	1.25
Cap, utility	2 @	.91
Cover, service cap	1 @	.73
Overcoat	1 @	37.00
Hankerchief	6 @	.18
Name tag	2 @	.30
Total		$260. 36

APPENDIX III

SOME OF THEIR SLANG AND WHAT IT MEANS

GI's have their own language. It is full of official and unofficial acronyms. They use shortcuts when talking. These are some of the more common terms used during the Cold War

AC&W – Official: Aircraft Control and Warning. Unofficial: All confusion and wonderment or Always Cleaning and Waxing or Always Crying and Whining.

AFCS – Official: Air Force Communications Squadron Unofficial: Alcohol First, Communications Second.

AFSC – Official: Air force Specialty Code.

APs – Official: Air Police. Unofficial: Apes, Sky Cop, Ramp rat.

ARMY GLOVES - In basic, having hands in your pockets.

B-52 – Official: Main bomber used by SAC. Unofficial: BUFF, Big Ugly Fat Fellow, et al.

BIRTH-CONTROL GLASSES - Military issue eyeglasses, the frames so ugly no one would have anything to do with you.

BLANKET PARTY - A blanket party or shower party was used when someone wouldn't take a shower or would not follow instructions. A blanket was thrown over him and he and the blanket were tossed into the shower and washed.

BOONDOCKERS - Brogans worn with the fatigue uniform.

BOOT - Trainee.

BROGAN MAINTENANCE - Fixing a problem with force usually a kick.

SH** FROM SHINEOLA – Being ignorant, as in can't tell sh** from Shine-o-la (a brand of shoe polish).

CIVIES- Civilian clothes.

COUNTING WAKE-UPS - The number of days before reassignment or separation, usually posted on a door or wall.

DILLIGAF - Do I Look Like I Give A F***.

DO YOU HAVE A NEED TO KNOW? - Used in Classified areas, or by your boss so he does not have to give you an answer to a question.

FIVE BY FIVE- Used to acknowledge a good strong radio or telephone signal being received.

FIGMO – F*** It, Got My Orders.

FLING FROM WING - Referring to instructions, real or imaginary, from higher headquarters.

FLYING ICE CREAM CONES – Instructor Badges worn at Keesler AFB Tech Schools.

FUBAR - "Fouled" Up Beyond All Recognition.

GET BENT – Bent meaning broken. As don't get Bent.

GI – Government Issue, meaning military uniforms, personnel, and equipment.

GI PARTY - Meaning a comprehensive top-to-bottom clean up, normally the barracks or work area.

GIG - Penalty for military infraction usually associated with inspections.

GIG LINE- Line formed by the leading edges of the shirt, belt buckle, and trouser fly when in uniform.

GMT - Greenwich Mean Time after the Greenwich meridian (0° Longitude).

GO TELL IT TO THE CHAPLAIN - Don't complain to me, or don't whine.

GREENBEAN'S - New arrivals, also newbies and Jeeps.

GUNG HO – Being too enthusiastic about the regulations.

HEAD UP THEIR ASS - Doing something really stupid. Not thinking at all.

HURRY UP AND WAIT – Rushing to get somewhere and then to stand around.

IHTFP - I Have Truly Found paradise, or I Hate This F***ing place.

JAFO - Just Another, F***ing Observer - Used by pilots to describe any non-Pilot.

JEEP - From the term used meaning General Purpose (GP) vehicle. Also a new recruit.

KP – Kitchen Police. KP duty is a helper in the mess hall cleaning pot and pans, peeling potatos scrubbing floors etc. It is a tiring, dirty job assigned to recruits and to lower ranking GI's.

LIFER - Career military person. Used derogatively meaning-Lazy Inefficient, Expecting Retirement.

LOWER THAN WHALE SH** - How low all recruits are according to TI's.

MAKE AN APPOINTMENT WITH YOUR CHAPLIN – Usual response given to a person whining.

NO SWEAT – Don't worry about the details.

NUGATORY - Of no real value, trifling or worthless.

OMGIF! - Translated as "OH my God, I'm F***ed!

PATS - Personnel awaiting tech school.

P. I. (Political Influence) - Notation found marked at the top of your records if you were acquainted with some high ranking politician or ranking officer.

PINGER - Referring to the short haircut received during basic training. According to TI's the short hair makes the "pinging" sound at night when the head rolls in the pillow.

RBS – Radar Bomb Scoring Units: Operated by the Strategic Air Command and used to score how close electronic bombs land to specific targets. Unofficial: Rakes, Brooms & Shovels.

RAINBOW – The varied colorful civilian clothing worn by recruits before being issued uniforms.

RHIP - Rank Has Its Privileges.

RHIR - Rank Has Its Responsibilities.

SAGE – Official: Semi Automatic Ground Environment. Unofficial: Soviet Aircraft Guaranteed Entry, Somebody's Always Getting Excited.

SAGE POPCORN - Aspirin.

SEALED BEAMS- A derogatory term used by TIs meaning eye glasses.

SEVEN LEVEL SCREWDRIVER - A special tool for specialized maintenance, to be utilized only by or with authority of a person who has achieved a high skill level in job proficiency. **R**

SHORT-TIMER, OR BEING SHORT - coming up quickly on a date for reassignment or separation.

SHOWER PARTIES - Forcibly delivering a fellow-airman, fully clothed, to the barracks shower to help celebrate a promotion or reassignment.

SKATER – A person who gets out of doing something which then is assigned to another.

SKIVVIES – Undershorts.

SNAFU - Situation Normal; All "Fouled" Up.

SOS- Creamed, chipped beef on toast; also   Sh** On a Shingle.

SPIT POLISH - Using water and shoe polish to shine boots and shoe.

SPIN AND POLISH - A person who is always in perfect uniform or one who follows military instructions to the letter.

T.I. – Training Instructor, usually in basic or advanced basic training.

THE POOP FROM GROUP - An expression referring to rumours.

TIGHT JAWS - Being angry enough to have visibly clenched teeth.

WTFO - What the f***, over. This is slang used to express astonishment.

USAF – Official: United States Air Force. Unofficial : Uncle Sam's Animal Farm or You Sure Are F****'d.

UTC- Coordinated Universal Time based on the Atomic clock and came about on JAN 1 1972.

ZULU TIME - Time referenced to GMT (Greenwich Mean Time).

APPENDIX IV

These are some of the Radar Sites Where They Worked and Lived and some individual comments:

Eufaula, AL

Joel Treshansky

We had inflatable radome covering the AN/FPS-26 at Eufaula, AL. One weekend, the radome sprung a leak in of all places at the very top center of the structure. I was sent (along with an airman named Manuel A. DelaRosa Archilla) to put a temporary patch over the leak. We built up the work platform, section by section, until it was high enough to reach the top. I had the dubious honor of crawling onto the highest platform, lying on my back, and applying the glued patch to the leaking edges. We then got the radome re-inflated, and all was well. (Funny how you will do crazy things at a young age that you would never even consider (doing) later in life...)

Tin City, AK

Bob Moore

I assure you there was little or no recreation there at this time (1962). Unless you want to consider movies where Joan Crawford was a teenager, shown two or three times a week, and dependant on if a airplane could land at the site. There were no bowling alleys, no television, AFN radio was sent to us via microwave, which we received at best over a lot of static. Our days consisted of getting up, going to work, getting off, going to the bar, drinking ourselves silly, going to bed and getting up the next day and doing it all over.

Middle-ton Island, Gulf of Alaska

Bob Slone

Located in the Gulf of Alaska, it was a horrible assignment for an 18 year old. It was isolated.

King Salmon' AK

Edward Franklin

I enjoyed the site even though remote. One could take leave and travel anywhere, including military hops. Since King Salmon had a commercial airport (USAF used the runways for jet fighters and transports) travel was easy, if the CO would sign a chit.

Cape Lisburne AK

James Howerton

Lisburne was a sight very few have experienced, bleak, treeless, a sea of ice clear to Russia, and 95 other guys. If I am correct, we were about 450 miles north of Nome and 230 miles northeast of Kotzebue.

Fire Island, AK

Pete (Not further identified)

I learned in this radar group, that my year on Fire Island was
like a vacation compared to a tour at Sparrevohn, and Indian
Mountain. I did not know that at the time. We were on a rock
with the lights of Anchorage in the distance. It was a constant
reminder of our life on the island and of other life in the
distance. A small taste of what it was like for the people on
Alcatraz Island (Federal Prison) in San Francisco Bay. But, we
were not caged on Fire Island and I am sure we had better
food. I still eat SOS - but NOT the chipped beef version - TOO
salty. I never run out of cans of Spam on the shelf - I buy the
25% less Sodium version now.

Yuma, AZ

Bob Wood

This had to be one of the worst! I spent 9 months there (1959-
60), and was glad to get orders for Okinawa, even if it was
10,000 miles from home. I grew up in South Texas, but I must
say that the heat in Yuma has Texas beat all over the place. We
had to put up with USMC MPs at the front gate too.

Winslow AZ

Jack Miller

Home of the 904th Radar Squadron and my home in 1962 and
63. Lived on base as I was on call as the NCOIC of Security
24-7 unless TDY someplace. We had 21 homes for married
NCO's and officers. Doc Pearsons wife ran the Exchange
annex and we built a great all ranks club. Many parties and
dances were held there. Great group of people who made
security an easy job. Relations with the civilian population and

the local police were excellent. It was not the greatest assignment but it beat the isolated and remote tours.

Texas Towers, Atlantic Ocean

Ron Henderson

I think the Texas Towers were about as remote as you could get. You sure didn't go anywhere until the chopper or the ship came out, usually 45 or 50 days.

Frobisher and Bay Baffin Island

Bill Seiter

Nothing to do. I was with the first group to occupy both sites. We had one ping pong table and that was it. There was no doctor which was scary as we were weathered in much of the time. However, fishing was good if you could stand the skeeters.

Frobisher, Bay

John Corbett

We were an early warning site with no Movements and Identification section or scramble capability. We were 32 miles inside the Arctic Circle and remote as all get out!

Puntzi Mt , BC Canada

Ralph Barrett

We were able to reach this base only by air. We had an L-20 aircraft to fly us in to our runway. We were about 100 air miles west of Williams Lake, B.C. We were actually close to nothing.

Mount Laguna, CA

Bill Stansbury

The tarantulas would gather by the hundreds at night on the outdoor screens of our heat exchangers because of the warmth. Loved the view of the desert, the Salton Sea, and beyond from the outside deck on the second level of the tower.

Ed Myshak

It was easily an hours drive up into the mountains from outside of San Diego to the site at Mt Laguna. On a clear night you could see Yuma, AZ and San Diego, CA. You could also see the Palomar Observatory on another mountain peak.

Mill Valley, CA

Tom Koselke

This place was great due to its location. It was in a state park 15 miles north of San Francisco.

Bethany Beach, DE

Larry Whitten

I was assigned to the gap filler section, where the towers were only 75 feet tall and had a stairway and not a vertical ladder. Also, all the equipment was in a building on the ground. It made the daily checks and Preventive Maintenance checks a lot easier and safer. I was stationed at the 771st Radar Sq. Cape Charles AFS, Va and traveled to DE.

Patrick AFB, FL

Dave Williams

One of the best assignments anywhere. The domes were right across highway A1A from the Atlantic ocean and the beach. The GATR site was sort of between the golf course and the runway. All the conveniences of a big base, all the pluses of being a detachment. I spent almost 3 years there and loved every minute.

Key West Naval Air Station, FL

David Semon

The radar site was on the north side of Boca Chica Key (Island) just east of the main street coming into the base which is at the 7 mile marker up Hwy 1 from the Court House (Mile Marker 0) on Key West Island. I lived on base in a dorm not far from their favorite runway made to look like a carrier deck and heard many full burner take-offs with F-14s and other aircraft.

Savannah GA

Jim Sanders

Our squadron commander Col. Quenin showed up unannounced in the 26 tower one afternoon with a major I had never seen before. They walked up to me and Col. Quenin said Major so-and-so has some questions about security he wants to ask you. After the first question my question to the commander was 'does the major have a need to know.' Col. Quenin smiled and said yes airman he does. So after passing the test, I answered his questions. We were a class A security area.

Mt Kaala, Oahu, HI

Jim McNamara

I was Tech Advisor to the 169th ACWS (Hawaii Air National Guard) from Nov 82 to Oct 86. The Mt Kaala guys were way better technicians on the height finder than I was ever going to be, so my technical advice wasn't particularly necessary up there. During my stint, the unit went from a manual operation to an automated system.

Hofn, Iceland

Bob Fletcher

By far the best assignment ever!

Rockville, Iceland

Paul Goldschmidt

I was at Rockville twice for a total of 4 years. They had a good club and great dining hall.

Cottonwood, ID

John Kwaczala

I did not especially like KP, but have to say at the 822nd, it was OK. We had a great cook there who always made steaks for us when I was on KP. Some of the best I have ever had. Better than Ruth Chris. I don't know if this had a bearing on it, but I did buy a pair of shoes from the mess Sarge who sold shoes as a side business. The 822nd was a great place to serve. We had our differences, but we were family.

Arlington Hts. IL

Jim Tarbet

Arlington Hts, IL - 755 ACWS, 1 mile SW of downtown, with major local streets. About 2 miles from the Northwest Tollway (I-90) and 1 mile from Illinois 53. We shared the site with an Army Nike unit. I know there were others that were not lucky to be on Gobbler's Knob, they could only get to their sites on the back of a mountain goat.

Waverly IA

Gene Oathout

A couple of years ago I drove through Waverly AFS Iowa and all buildings appeared to be intact except for the entrance guard shack. Radomes and sails were gone and the towers seemed to be in use by someone. I recently had an inquiry about base history from someone who just occupied one of the on-base houses that were falling apart and overgrown with weeds when I visited. Apparently Habitat For Humanity is in there working those houses.

Yoder, KS

Bob Moore

I was in the 793rd AC&W at Hutchinson during the Cuban Missile Crisis. We wondered what that had to do to us in the middle of a wheat field in the middle of Kansas. None the less, we spent days on end filling sand bags and stacking them all around the operations building to ward off any stray missiles that might have found their way to us.

Ft Knox, KY

Thomas Barnes

During my assignment to the 784th RADRON, Snow Mountain, KY from 25 Oct 65 to 8 May 68 we had combat exercises regularly. Many of these were boring, but some of them got to be fun. At no time were we told that Air Defense Command radar sites were target #1 for any attacking enemy. My fellow airmen and I always thought we did the sabotage exercises for the magic term, "general military purposes."

Saglek (Pine Tree Line), Labrador

Fred Boutin

It was a highlight in my life. Exciting, interesting, and always surprising me. The cook would bake fresh bread in the middle of the night when we were on duty in the tower and we would sneak down and devour fresh bread with a half-pound of butter and sometimes a rasher of bacon. It paid to be nice to the cook. We were there doing what we learned to do up on our cliffs and that was to read all the postings or anything else. It was either read or get drunk. There wasn't much else to do. We did try though. Everyone pulled together to do what we could. Someone ran the movies, someone else ran the BX someone else ran the two-lane bowling alley and so on. I ran the photo lab and special services which didn't amount to much when I got there and not a great deal when I left. Too hard to get equipment and keep it in shape. I'm not hitting the grapes sour or otherwise but it could have been better equipped but we still had fun.

Houma, LA

Paul Geckeler

When I was at 657th AFS Houma, LA in '69 I tried Dixie Beer. The reason I tried it was they had what I thought were the best commercials. Gentle music with an old southern mansion behind the Spanish Moss. After seeing the commercials I figured it had to be a good beer. They just had a good advertising agency.

Fort Williams ME

Bob Cumby

A lot of you don't remember the old AN/TPS-1B Lashup sites back in 1950. I was at one in 1951 at Fort Williams, Maine. The Portland Head Lighthouse is there. Locals refer to it as the Head Light. The TIPS-1B was near the Head Light. There may been other old Lashup sites near the sea at that time. Fort Williams is now a state park. If you are ever in the Portland, ME area you should go there. It one of the most beautiful parks. The Portland Head Light is the most photographed lighthouse in the world.

Ft Meade, MD

Paul Ries

It would be good to keep in mind that the SAGE system was used by the Army Air Defense Command in cooperation with NORAD. I worked in the AADCP located at Ft. George G. Meade in MD. We were associated with the 770th AF unit also co-stationed with us along with the Navy and worked together on all aspects of the coastal defense system. Nike Hercules was just one component of that defense system and was tied through the AADCP to the SAGE operation at Ft. Lee Virginia.

North Truro, MA

Lon Molnar

Yes I remember one very nice part of the recreational life at North Truro AFS. We held the Air Division Softball tournament there in 78 and 79. Between games it was fun to head out to the beach, climbing down what seemed like a very steep cliff to get to the water. But at the bottom was quite a treat...it was a nude beach! That left quit an impression on me!!!!

Battle Creek, MI

Dave Ross

When leaving Det33 Kamo, Japan I received orders for my next assignment to Custer AFS, Michigan. I got a Michigan Map and found Custer, Michigan on it. I Left Indianapolis, IN. for Custer, MI. It was a beautiful drive up the East side of Michigan past Ludington, Holland then to Custer. After arriving I didn't see anything concerning an Air Force Station.

I stopped at a little Post Office, and asked the older guy running it where I could find Custer AFS. He kind of laughed and told me I was about 150 miles to far to the North. I needed to be at Custer AFS in Battle Creek, MI. After the look on my face he said "Son you're not the only one that made that mistake."

Sault Ste. Marie, MI

Gordy Stiles

We definitely had a drill team, and a good one at that. I was at "The Soo" from Mar '62, until Nov '64, and had lots of fun doing the drilling. It was under the guidance of SSgt. Russell Parker. We had about 15 to 20 marchers, I don't remember how many. We marched in all the parades in downtown Sault Ste. Marie, and all the ones across the river in Sault Ste. Marie, Ontario (those bagpipes were impossible to march to), and always marched in the Labor Day parade at St. Ignace, MI. Lots of fun, lots of memories.

Calumet, MI

Jack Armstrong

The T-33 (jet aircraft) was there when I arrived in Sept 1974 and still there when I left in May 1976. It was on a pedestal on the North side of the road just after the last curve and before the entrance to base housing. At the turn-off to the site from US 41, someone had built a giant thermometer. It was graduated in feet not degrees. It was used to measure the snow depth. I never saw it go above the twenty-nine foot mark.

Crystal Springs MS

Gerry Heaton

I was in from '61-'81. The class "A" blues I was issued had the great big "patch" pockets which were later changed to inside pockets with a flap, and the breast pockets like the army's. To me that was and always will be the best looking uniform the Air Force has ever had. Thank goodness they are getting away from the latest "business suit" fiasco and back to a more "military" style.

Biloxi, MS

Steve Weatherly

I passed through Keesler AFB Annex 1 and 3 while attending the Ground Electronics Officer course from May 64 to Apr 65. Initially we attended electronics fundamental classes at Keesler AFB Annex 3 down Hwy 90 in Gulfport. Our class finished up at Annex 1 for specialized electronics training where we dealt mainly with the FPS-20 and FPS-6 radar. Unfortunately the FPS-6 was down the entire time.

Fordland, MO

William Bacon Sr.

The old site of the 746 AC&W Sq at Fordland Missouri is now a State prison. I visited this site this in 2007. I was stationed there 54 to 58. Some of the old buildings are still in use. The Search Radar Tower (without the bubble) is used and the power plant is still in use with the same equipment. The Ops Building and several of the barracks are in use today. The Roses that we planted on the fence are gone. The chapel we built is gone, however several of the base houses are still there and in use.

Yaak, MT

Dick Spahl

I was stationed at the 680th Yaak, Montana in winter of 54-55. We had 8 ft. of snow in March and were cut off from the outside world for almost a month. Had to be supplied by airdrop.

Kalispell MT

Ed Lupien

The site was near a number of lovely small towns including Kalispell, Whitefish, and Columbia Falls. The Lower Site was about a mile from the little town of Lakeside on Flathead Lake. The Rocky Mountains and lakes are beautiful. Glacier National Park is just east of Kalispell. We used to go snow skiing on Big Mountain near Whitefish. There was water skiing on Flathead and Whitefish Lakes.

Opheim MT

Lyndon Lynch

We lived in housing at the old Glasgow AFB in Glasgow, and took the "blue goose" (USAF bus) to work and back every day. This worked out ok, except when it snowed so much that the only way we could get there was to follow the plow, as you couldn't see the road.

Cut Bank MT

Ray Sturgill

I was at the 681st AC&W Radar Site in Cut Bank, MT Jan. 1957 thru Jan. 1960. I wish I could locate a Site Patch for the 681st.

Miles City, MT

Robert J.F. Peru

I thoroughly enjoyed Miles City and the squadron. We had excellent hunting, very friendly natives.

Lewistown, MT

Tom Whiteley

A civilian contractor had been hired to install two large water towers at the site. While digging his crew found gold veins in each of the two areas dug. One was 18" and the other 36" wide. The contractor took out the mineral rights as the Air Force did not possess them. I took home some samples of the gold and my father in law, who had been a mining engineer in Peru, said it was very rich. I also took home some black diamonds, which rock hounds love.

Beausejour, Pine Tree Line, NB, Canada

Charles Wayne Fitzgerald

I was at Beausejour, and despite all the hardships and harsh environment, I loved it.

Omaha, NE

Gerald Bjerke

I went back to Omaha, Neb. and visited the old 789th site. Most of the buildings and housing were still intact. The radar was being used by the FAA. A Union group had offices in one of the newer buildings. I was able to drive in and around the old base. I talked to union office first.

Hastings NE

Brian Dorney

A huge naval ammunition depot in the middle of Nebraska was the site of the 625th Radron, which the AF built in the early 60's. The Navy vacated and turned the site over to AF control in May '66. There was also an AF Radar Bomb Scoring group on site.

Fallon, NV

L.A. McClurg

Fallon home of the 858th AC&W started up in 1956 thru the mid 70's. I spent all 4 years in Fallon, it was a great place to be at. In Fallon you were 60 miles from everywhere; Carson City, Reno, Lovelock, and Yerrington. We had a great group of guys back then. It was a Naval Air Station with AF radar. We even had our own newspaper called "The Warners."

Winnemucca, NV

Art Leighton

Had a very nice swimming pool, and in the summer heat, was very popular by all. I once got my daughter's pony a little tipsy, by giving him beer, and he just loved it. Of course my daughter wasn't too happy about it as she had to walk the pony home. The event was a Squadron Poolside Picnic in 1964.

Highlands, NJ

John Corbett

I spent 18 months there starting in Jan 58. The duty was great. We were a prime direction center and scrambled out of McGuire AFB NJ and Suffolk AFB NY. The local was on the North Jersey coast and I could walk to the beach in Sea Bright. Now I live less than 8 miles from there.

Palermo NJ

Ken Zwizanski

Many sites were situated on or near the coast. I was at Palermo and I bet it wasn't more than 10 or 15 feet above sea level. I'm sure North Truro and Cape Charles and Montauk and many others were about the same.

Moriarty, NM

Darrel Pittman

My first site right out of Keesler was at Moriarty NM in 1955. It was a great place and the AF members there were very hospitable to everybody. As a close knit group we did lots of things together, Officers and enlisted alike.

West Mesa AFS, NM

Charles Glover

Does any one know why the military did away with Fast Pitch Softball? I really enjoyed those days. I pitched for the 687th Radar sq., West Mesa AFS in Albuquerque in 1963 and we won the Championship at Kirtland AFB. We also played AAA in the Albuquerque area. I really enjoyed and miss those days.

Schenectady, NY

Bobby Johnson

If there was any better site to be stationed at in the world, I would like to know about it! That was an experience for this shy southern teenager. Met some wonderful people there that almost sent my life in another direction, but I did not go back as I promised. While in Japan with 10th Radar traveled to about 15 different sites in Japan and Korea; most were isolated, cold, windy, lonely, and all of them on the highest mountain in the area.

Saratoga Springs NY

Jim Sanders

I contacted the owner to access the 656th site in Saratoga Springs, NY. I was there in 2007 and just happened to catch a lucky break that he was in town from CA during late summer. I just wanted to add this cautionary advisory to any so inclined that you don't travel up the mountain expecting to simply drive on in. There is (or was at the time) a local caretaker or property manager based in Saratoga Springs, but he is not empowered to admit "tourists" to the site.

St. Anthony, (Pine Tree Line), NFdl, CN

Cliff d'Autremont

The town of St. Anthony, population 2000, was 2 miles down the hill and even had a local hospital. On Saturday nights some of the girls from the hospital would be our guests at the Airmen's and NCO clubs. So, even though we were remote, we weren't "that" remote. What made me feel remote was that the GIs weren't allowed to have cars. But there weren't many roads anyway. Our gravel road to the site that was maintained by the USAF was said to have been the best road in northern Newfoundland. The roads there are all paved now.

Roanoke rapids, NC

Wayne Williams

I was stationed at the 632nd twice. I visit Roanoke Rapids quite frequently as my wife is from there. Most of the buildings are still standing, the AN/FPS-26 and -27 towers are there but the FPS-6 tower was taken down. The site is in a deplorable condition and is privately owned. I go to it every now and then. It's best to visit in the winter when the snakes and spiders thin out.

Fortuna, ND

Bill Miller

The wind indicator at the 780th went to 110 knots, 126 MPH. In winter I have seen the indicator pegged for many minutes at a time. The FPS-6 tower door opened to the north, and lots of times getting out of the tower involved leaning on the door as hard as we could, waiting for the wind to drop a little. Finally, they built a wooden wind-break on the north and west sides which helped a lot. When I first got there, there were little cedar or pine trees planted around the base. They were about 4 feet tall. I think they got shorter each winter.

Minot AFS, ND

Gary Smith

It didn't take long for me to get fed up with winter and I had three of them before I could get away. Minot and NoDaks were good people, friendly, hard-working. Remember the two coolies between town and the site? A devil of a drive in winter.

Bellefontaine, OH

Robert Dublin

In May 68, I was working evening shift in the 26 tower. It had a 40 ft. uncovered extension and at the base of the tower was a diesel fueled emergency power generator that supplied electrical power for the FPS 6, 26, and 27 Radomes. We lost power about 6PM and the power technician was having trouble getting the generators on line. I went to start the radome generator. It would not start and I cranked it until the battery was dead. I was really concerned as the search radar (FPS 27) antenna as it was up on blocks having bull gear work done. After 2 hours they got the generators online. By then the FPS 27 tower bubble looked like a mushroom but didn't touch the antenna. I'm glad that only happened to me once.

Bartlesville, OK

Bob Burns

I was at the 796th AC&W Squadron at Bartlesville, Oklahoma. There is not much left of the site west of Bartlesville. My last visit was several years ago and it was occupied by a Candle Manufacturing company. You could drive up the hill to (the) site. Some of the housing was still there, but not in very good shape.

Oklahoma City, OK

Leon Gall

I was stationed at the 746th AC&W Squadron, Oklahoma City AFS in 1964 when the HQ Battery for the Nike Missile Defense at Dallas TX was moved from Dallas to our location. The Army buildings all had rocks painted white on either side of the sidewalks leading to them and the soldiers mowed the grass and trimmed the hedges at their barracks buildings. The Air Force buildings had no "white rocks" and the grass and bushes were cut by the Base Civil Engineers. I guess they are right. The Air force is the gentleman's service.

Armstrong AS, Pine tree line, Ont, CN

Ken Jones

If you had your family with you, it was considered to be an accompanied tour, but if you were young and single, as most of us were, it was a remote tour. If you loved to hunt and fish, as I did, however, you were in Hog Heaven. It was the only assignment I have heard of where part of your in-processing was to assign you a barracks, a room, a mess pass and a local fishing license free.

Mt. Hebo, OR

287

Gary Smith

After 18 months in the Philippines I was not ready for the Hebo winter. The tunnels should have told me something but it kind of crept up on me. When winter did hit there was no mistaking, we were in the deep doo-doo. Thank god for tunnels and ropes and the BUS. Oregon has some of the most beautiful country ever seen and the trip was worth it.

Benton PA

Richard Doty

We went into Williamsport or Scranton for the nightlife.

Oakdale PA

Larry Smith

I was an Airman Second Class when I got to the 662nd Radar Squadron (SAGE) (ADC) at the Oakdale Army Installation, Oakdale, PA. The Air Force squadron was a tenant on the Army's Missile Master site. We lived under Army rules but worked under Air Force rules.

Wallace AS Philippines

Jim Truman

I liked it so much I stayed for three years. It was on the beach, had great people and the food at the club was the best. The San Miguel beer was always cold on and off base; who would want to be anywhere else.

Charleston, SC

Dan Dawdy

We always had good food at the 792nd. Troops from nearby Charleston AFB and the USN base came by regularly to (enjoy) our steak and fries day.

Aiken, SC

Don Skinner

There was a 861st fallout shelter in '68 when I got there. Some of us were picked for training and spent a week in the place, with canned water, C-ration crackers, and 10 whining people. But I still have my Shelter Manager certificate!

Gettysburg, SD

Raymond Vena

Life at a remote site was the best there was. The people in the community were great and to this day 52 yrs later I still correspond with friends there. I attended the 50th reunion and 120 guys and wives showed up. We had a ball. I was there in 56 and 57. The site today is now a ghost town with its Quonset huts falling down and grown over with prairie grass. My kids couldn't get over seeing the place. Those were the days.

P-Y-Do Island, Korea

Jim Brown

Went TDY to P-Y-Do in 64 or 65, don't remember. I do remember seeing the telephone poles on wheels made up to look like heavy artillery. Big excitement when I was there was a Mule Train truck with six drive wheels coming off a landing craft as the tide was coming in. It broke down and mired in the soft sand. Later all that could be seen was the top of the cab. I had a very memorable ride in a C-47. Ice was being thrown off the wings and props. Ah, the memories.

Kang-nung , Korea

Richard Waddell

During one especially heavy snow, when the roads were closed, we ran out of our usual food and ate C-rations for a few days. At the same time, we also ran low on oil to heat our tents and Quonset huts. C-119 crews dropped heating oil to us by parachute. I have photos of parachutes with oil drums trailing out the back of a C-119. Before the next winter, they brought in a supply of oil using an LST. (1956)

Madrid, Spain

Ben MacDonald

It was my first assignment directly out of Keesler AC&W School in the summer of 68. I stayed 5 years and they had to drag me away kicking and screaming to head back to the States.

Bradyville, TN

Tom Page

This former radar facility is now a garage and a home to hogs and chickens. The building and the fence are mostly intact; however, the radar-tower foundations have been removed.

Zapata, TX

Ed Murillo

In 58 the 742nd was a brand new site located on the Rio Grande about 55 miles south of Laredo TX. Our Main job was Surveillance and M&I we also provide assistance to the many USAF student pilots from Sheppard AFB and Navy Pilots out of Corpus Christie TX. We were a check -point for many Airlines entering the US from Mexico. Most of the traffic in our area was low and slow aircraft crossing the border from Mexico, some we could ID, for the others we usually called the Customs or Border Patrol and notified them of possible landing sites.

St. Albans, VT

Bill Elliott

Good duty with excellent supervisors and people. The converted Park Ranger's residence made an excellent dining hall and outstanding club. Close to town with people who welcomed GI's, close to Lake Champlain, and plenty of good fishing and hunting.

Apple Orchard Mt. VA

Bob Felberg

The 649th AC&W Squadron was located at Bedford AFS on top of Apple Orchard Mountain, the highest point in VA I Believe. Two roads led up to the Blue Ridge site. One was the Blue Ridge Parkway which was restricted pretty much to automobile/tourist traffic. All trucks needing to get to the site had to use a rugged access road going up the mountain from the western side of the Blue Ridge. There was a herd of around 30 elk that hung around in a meadow by the Peaks of Otter (entrance to the Blue Ridge Parkway from Bedford). The elk were shipped off to somewhere out west and there is a lake and hotel at that spot now.

Cape Charles, VA

Larry D. Whitten

I was stationed at Cape Charles AFS, VA twice. My first tour, we had to use the ferry boat to get to the GF (Gap Filler radar) site in North Carolina and support base at Langley AFB, VA across the Chesapeake Bay. Good chow at the mess hall, especially on Tuesdays and Fridays. Tuesdays was steak with all the trimmings and Friday was a seafood smorgasbord. Deeeeeeeeeelicious!! Second tour was almost as nice. No Gap Filler in North Carolina this time and no ferry boat ride. Now there was the Chesapeake Bay Bridge and Tunnel. We were able to get across the bay in about a 1/2 hour to get to Langley AFB.

Curlew WA

Dan Wilson

You started your shift with a nice 45 minute ride up the mountain dirt road, freezing in winter and sweating during the few weeks of spring/summer/fall, always under a tarp in the back of a six-by, which sucked in the clouds of dirt and covered you completely.

Neah Bay , WA

John Kimmes

Beautiful in the summer but cold, windy and
rainy during the winter months. One blacktop road got you out
of the area. It wasn't uncommon during the rainy season to
have that road blocked with boulders and dirt due to mini
landslides. The only way out then was a dirt logging road.

Tom Goodman, added this about Neah Bay:

Pretty remote for a stateside tour. The closest "big" city I can
remember was Port Angeles, WA.

Othello, WA

Larry Clare

When I left Othello in 1961 all of the area outside of the fence
was nothing but sagebrush except the base housing to the east.
The only way for trees to grow would be to pump water to
them. Another five miles to the west marked the boundary of
the Hanford Atomic Energy Area.

Colville , WA

Stan Anderson

The bubble checks we would get from the F-102 unit at Geiger
were memorable. There was a lake south-east of the site over
500 feet below topside. One time an F-102 jock came in low
over the lake and up the side of the mountain mighty close to
tree-top level. It was an awesome sight and experience to stand
on top of the mountain and see a jet fighter flying up toward
you. I'll never forget it. Of all the so-called bubble checks, it
was the greatest.

Thomas, WV

Larry Smith

During 1965 I was assigned to Gap Filler Radar Maintenance out of the main host radar station at Oakdale Army Installation PA where the 662nd Radar Squadron was located. We had three Gap Filler sites to maintain; Thompson, Ohio; Brookfield, Ohio, and Thomas, West Virginia. My crew drove into West Virginia and stayed 3 days. Remote control of the site was handled back in Oakdale. If there were maintenance problems that the Operations people could not reset, they would call our stand-by team to come out, check out the equipment at the main base and maybe have to travel to WV for emergency maintenance.

Williams Bay , WI

Ed O'Neill

There were four small towns within 15 minutes. There were many WWII vets in these towns and they treated us wonderfully, and the girls were attractive and friendly. It was, and is, a strong tourist destination and during the summers the whole area was awash with pretty young college girls, teachers, nurses, secretaries, etc.

Marty Fisher added this about Williams:

It was a great duty station. Lake Geneva had many of named bands play there on the lake. Down the road to Delevan Lake there were lots of activities. I have nothing but great memories of the site and area.

Jack Miller